Practical Finite Element Simulations with SOLIDWORKS 2022

An illustrated guide to performing static analysis with SOLIDWORKS Simulation

Khameel B. Mustapha

BIRMINGHAM—MUMBAI

Practical Finite Element Simulations with SOLIDWORKS 2022

Copyright © 2022 Packt Publishing

Associate Group Product Manager: Rohit Rajkumar
Publishing Product Manager: Aaron Tanna
Senior Editor: Keagan Carneiro
Content Development Editor: Adrija Mitra
Technical Editor: Joseph Aloocaran
Copy Editor: Safis Editing
Project Coordinator: Rashika Ba
Proofreader: Safis Editing
Indexer: Pratik Shirodkar
Production Designer: Roshan Kawale
Marketing Coordinator: Elizabeth Varghese

First published: February 2022

Production reference: 1301221

Published by Packt Publishing Ltd.
Livery Place
35 Livery Street
Birmingham
B3 2PB, UK.

978-1-80181-992-3

www.packt.com

To my parents, Hajj and Hajjah Muibideen Mustapha Balogun.
Your blessings and prayers comprise the engine that propels me
forward. To my wife, Aminah, for her spiritual support, psychological
companionship, and emotional connections.

- Khameel Bayo Mustapha

Contributors

About the author

Khameel B. Mustapha obtained his doctorate from Nanyang Technological University (Singapore) with a focus on the development of computational methods in the area of micro-continuum theory. He has years of experience working on a variety of finite element analysis platforms. Khameel has provided training to thousands of students and participants on the applications of finite element simulations to the analysis, design, and performance assessments of engineering components. His primary research interest is geared toward the mechanics and modeling of microscale structures, but his wider interest encompasses computational mechanics, engineering teaching philosophy, and mechanics of advanced systems (functionally graded materials, sandwich composites, subcellular biological structures, energy materials, and additively manufactured components). He is currently an Associate Professor with the University of Nottingham Malaysia Campus and has previously held a faculty position with the Swinburne University of Technology (Sarawak). He is a **Fellow of the Higher Education Academy** (**FHEA**), UK.

I wish to thank the developers of SOLIDWORKS software. This book would not have been possible without the existence of this brilliant piece of engineering marvel.

I acknowledge the institutional support of the University of Nottingham Malaysia Campus. Notably, I would like to thank the hardworking librarians for their persistent efforts in ensuring that the university's library collection is rich, diverse, and up to date.

I treasure the love and care of my wife, Aminah, and our children, Hibatullah, Abdul Alim, and Hameedah. I express gratitude for your unyielding endurance and endless patience in the sacrifices required to complete this book. Finally, all praise is due to the Owner of The Most Beautiful Names. He is the source of all knowledge and The Most Exalted.

About the reviewer

Paul Anthony has been a working mechanical engineer for over 10 years. He has spent his time practicing his skills in the automotive, solar, radiation monitoring, grinding technologies, machine spindles, and automation equipment fields. Analysis has always been a joy for Paul; it is a great tool for tying both the practical and design worlds together. Many applications for FEA have presented themselves during the course of his career, spanning simple static analyses to much more complex vibration and heat transfer problems. He looks forward to the future and all the opportunities that these skills will present to him.

Paul would like to dedicate his efforts and extend his gratitude to his Aunt Melissa Swenson and Uncle Scott Swenson for an influential conversation they had just before he embarked on his engineering career. Their words have stayed imprinted on his mind to this day, and they cannot begin to understand how grateful he is to these two wonderful people for their influence on the course of his career and life.

Table of Contents

3
Analyses of Beams and Frames

4
Analyses of Torsionally Loaded Components

Section 2: SOLIDWORKS Simulation with Shell and Solid Elements

5
Analyses of Axisymmetric Bodies

6
Analysis of Components with Solid Elements

7

Analyses of Components with Mixed Elements

Section 3: Advanced SOLIDWORKS Simulation with Complex Material and Loading Behavior

8

Simulation of Components with Composite Materials

9

Simulation of Components under Thermo-Mechanical and Cyclic Loads

10

A Guide to Meshing in SOLIDWORKS

Preface

SOLIDWORKS is a ubiquitous software at the forefront of technologies for the three-dimensional modeling and designing of components. SOLIDWORKS Simulation, which is the focus of this book, harnesses the power of finite element simulations within the robust SOLIDWORKS simulation environment. Predominantly deployed for detailed assessments of product performance, SOLIDWORKS Simulation has grown in prominence in recent years among analysts and engineers tasked with taking product designs to a whole new level through the seamless integration of virtual prototyping, performance diagnostics, and failure analyses.

As a subset of computer-aided engineering skills, finite element simulation was once delegated to specialists within engineering firms. However, as the line between engineering analysts and designers blurs with the proliferation of software such as SOLIDWORKS, many engineers are now required to be both familiar and proficient with complex engineering analysis related to performance evaluations of products. In this vein, learning SOLIDWORKS Simulation will significantly enhance your ability to contribute to bringing products to market faster. This book offers a path for you to acquire a foundation in practical **finite element analysis (FEA)** using SOLIDWORKS Simulation.

Who this book is for

This book is for professionals working in various engineering fields in which finite element simulation is used heavily. This includes engineers and analysts from areas such as the aerospace, mechanical, civil, and mechatronics engineering fields who are looking to explore the simulation capabilities of SOLIDWORKS. No prior familiarity with the SOLIDWORKS simulation environment is assumed. However, basic knowledge of modeling in SOLIDWORKS or any CAD software is assumed.

What this book covers

The book covers static analyses of engineering systems. Overall, the book provides several cases of static problems, presents a systematic illustration of their solutions, and describes the interpretation of the solutions from the perspective of an engineering analyst. A summary of the chapters is provided next.

Chapter 1, Getting Started with Finite Element Simulation, offers an overview of the **finite element method** (**FEM**) and highlights the uniqueness of SOLIDWORKS Simulation regarding the analysis of engineering components. You will learn about the SOLIDWORKS simulation interface, be introduced to the general steps for the simulation of single and assembly-based components, and understand the major families of elements.

Chapter 2, Analyses of Bars and Trusses, commences the proper exploration journey with a problem concerning the static analysis of a crane. It provides an introduction to the use of the weldments tool and explains how to change a beam element into a truss element.

Chapter 3, Analyses of Beams and Frames, focuses on the simulation procedures and strategies for the analysis of transversely loaded members in the form of beams. It expands on the knowledge of SOLIDWORKS's weldment tool and provides a strategy for the application of more complex types of loads (such as concentrated force, moment, and distributed load). It also demonstrates the idea of using critical points along the length of beams to create appropriate line segments.

Chapter 4, Analyses of Torsionally Loaded Components, deals with the static analysis of torsionally loaded members. It showcases the creation of our first custom material and highlights the extraction of the angle of twists following the application of torsional loads.

Chapter 5, Analyses of Axisymmetric Bodies, initiates the treatment of advanced elements. It discusses the attributes of shell and axisymmetric plane elements and applies these elements to two case studies in the form of pressure vessels and a flywheel. Via these examples, you will learn how to apply symmetric boundary conditions, duplicate studies, take advantage of the probe tool, and get exposed to methods of visualizing the 3D plot for a study conducted with an axisymmetric plane element.

Chapter 6, Analyses of Components with Solid Elements, examines the deployment of solid elements and showcases the analyses of helical springs and spur gears. Through the examples provided in this chapter, you will learn about mesh control, an assessment of contact stress, how to set up "no penetration contact," and become familiar with the significance of curvature-based mesh.

Chapter 7, Analyses of Components with Mixed Elements, brings together the major families of elements for the analysis of a multi-story building. This chapter explores the use of the automatic contact pairs detection tool. It also highlights how to deploy the in-built soft spring feature within SOLIDWORKS Simulation to provide stability for non-linear simulation studies.

Chapter 8, Simulation of Components with Composite Materials, taps into SOLIDWORKS's simulation capability for the analysis of components made up of composite materials. Among other things, you will learn how to convert a basic surface body into a composite shell, walk through the procedure to create a custom orthotropic composite material, and assign its properties to a composite laminate.

Chapter 9, Simulation of Components under Thermo-Mechanical and Cyclic Loads, is dedicated to analyses that involve thermal and cyclic loads. It discusses the integration of thermal and static analyses to address the simulation of components at elevated temperatures. You will also learn how to conduct fatigue analysis and get exposure to the optimization capability of SOLIDWORKS Simulation regarding designing components against failure.

Chapter 10, A Guide to Meshing in SOLIDWORKS, culminates in a brief coverage of methods to customize the meshing of structures to achieve reliable results. You will encounter mesh control for different types of elements and learn how to employ convergence analysis to evaluate the accuracy of simulation results.

To get the most out of this book

You will need to have access to a version of SOLIDWORKS that permits the use of SOLIDWORKS Simulation for all chapters.

To benefit from the book's hands-on approach to learning SOLIDWORKS Simulation, you should follow the step-by-step guidelines on various aspects of the simulation workflow as you read.

Further, to get the most out of the book, a basic familiarity with 3D modeling techniques using SOLIDWORKS, even if at a beginner's level, will help you to move along the chapters smoothly. Along with the aforementioned, it is assumed that you have acquired the essentials of elementary mechanics.

Software/hardware covered in the book	Operating system requirements
SOLIDWORKS/SOLIDWORKS Simulation	Windows

SOLIDWORKS Corporation offers three types of licenses for SOLIDWORKS Simulation. You will need the premium license to be able to work through all the chapters without any restrictions.

Download the example code files

You can download the example files for this book from GitHub at `https://github.com/PacktPublishing/Practical-Finite-Element-Simulations-with-SOLIDWORKS-2022`. If there's an update to the simulation models/files, it will be updated in the GitHub repository.

We also have other code bundles from our rich catalog of books and videos available at `https://github.com/PacktPublishing/`. Check them out!

Download the color images

We also provide a PDF file that has color images of the screenshots and diagrams used in this book. You can download it here: `https://static.packt-cdn.com/downloads/9781801819923_ColorImages.pdf`.

Conventions used

There are a number of text conventions used throughout this book.

`Code in text`: Indicates code words in the text, database table names, folder names, filenames, file extensions, pathnames, dummy URLs, user input, and Twitter handles. Here is an example: "You should check to see that it comprises a SOLIDWORKS part file named `Diaphragm`."

Bold: Indicates a new term, an important word, or words that you see on screen. For instance, words in menus or dialog boxes appear in **bold**. Here is an example: "With the **Simulation** tab active, create a new study by clicking on **New Study**."

> **Tips or Important Notes**
> Appear like this.

Get in touch

Feedback from our readers is always welcome.

General feedback: If you have questions about any aspect of this book, email us at customercare@packtpub.com and mention the book title in the subject of your message.

Errata: Although we have taken every care to ensure the accuracy of our content, mistakes do happen. If you have found a mistake in this book, we would be grateful if you would report this to us. Please visit www.packtpub.com/support/errata and fill in the form.

Piracy: If you come across any illegal copies of our works in any form on the internet, we would be grateful if you would provide us with the location address or website name. Please contact us at copyright@packt.com with a link to the material.

If you are interested in becoming an author: If there is a topic that you have expertise in and you are interested in either writing or contributing to a book, please visit authors.packtpub.com.

Share Your Thoughts

Once you've read *Practical Finite Element Simulations with SOLIDWORKS 2022*, we'd love to hear your thoughts! Scan the QR code below to go straight to the Amazon review page for this book and share your feedback.

https://packt.link/r/1801819920

Your review is important to us and the tech community and will help us make sure we're delivering excellent quality content.

Section 1: An Introduction to SOLIDWORKS Simulation

A journey of a thousand miles begins with a single step. In this same vein, this first section of the book introduces you to the interface of SOLIDWORKS Simulation and unveils how to conduct basic analyses of engineering components. By the end of this section, you will have acquired knowledge of simulation with one-dimensional elements in the SOLIDWORKS simulation library.

This section comprises the following chapters:

- *Chapter 1, Getting Started with Finite Element Simulation*
- *Chapter 2, Analyses of Bars and Trusses*
- *Chapter 3, Analyses of Beams and Frames*
- *Chapter 4, Analyses of Torsionally Loaded Components*

1
Getting Started with Finite Element Simulation

This chapter lays the groundwork for what is to come in the later chapters. It gives a high-level discussion of the **finite element method** (**FEM**), tracing its historical origin, emphasizing its application, and outlining its implementations. Some of the key concepts regarding FEM (such as discretization, types of elements in FEM, nodes, and more) are briefly highlighted, and a snapshot of the SOLIDWORKS Simulation interface, license, and computing requirements are discussed. To this end, this chapter covers the following major topics:

- An overview of finite element simulation
- Understanding SOLIDWORKS Simulation
- Getting started with the SOLIDWORKS interfaces
- What is new in SOLIDWORKS Simulation 2021–2022?

Technical requirements

You will need to have access to SOLIDWORKS software with a SOLIDWORKS Simulation license.

You can find the supporting files for this chapter here: `https://github.com/PacktPublishing/Practical-Finite-Element-Simulations-with-SOLIDWORKS-2022/tree/main/Chapter01`

An overview of finite element simulation

This section offers a short account of the historical origin, importance, application, and implementation of finite element simulation.

Background

The pervasiveness of **computer-aided engineering** (**CAE**) has grown in parallel with the progress in the development of digital computers. Historically, CAE was predominantly used for the solid and surface modeling of engineering parts and assembly. However, in recent years, a glaring inroad of this progress has manifested in the simulation of various forms of engineering systems. Indeed, simulation is at the heart of the progress for advanced product development across different industries. Specifically, in the area of engineering product development, finite element simulation, which is based on the rich theoretical framework provided by the **finite element analysis** (**FEA**), represents a crucial toolkit for the following:

- A smarter and efficient design of engineering systems throughout the product development life cycle

- Minimizing product recall through rigorous analyses and examinations of the fidelity of product performance

- Facilitating iterations of virtual prototypes before incurring the cost of building physical prototypes

> **Information**
>
> Simulation is a word that has many definitions. Its use in this book orients toward its definition as the representation of a real physical system with a virtual prototype to study, analyze, and predict its response under external effects.

These days, FEM can be regarded as a standalone subfield of activity within the larger CAE. However, historical records place its root in the field of applied mathematics. The first documented application of the method is linked to the technical attempt to solve design problems from the aerospace industry in the 1950s. Nonetheless, the method rose to fame in the 1960s with the work of *Clough [1]* (please refer to the *Further reading* section) and the publication of the first book on FEA by *Zienkiewicz* and *Cheung [2]*. Since these events, FEM has recorded many successes, and there has been an upswing of applications spreading from the automotive, aerospace, biomedical, civil, consumer products, nuclear, and mechanical fields, to the space industry.

FEA entails transforming physical processes/products into some approximate mathematical equivalents called mathematical models. Afterward, the models are solved with appropriate computing resources via numerical solution algorithms. Now, the notion of approximation here might evoke a feeling of inferiority of finite element simulation. However, it does not in any way detract from the excellent accomplishments of FEA, some of which will be demonstrated in this book. By the virtue of their complexities, most physical objects or practical products cannot be reduced to perfect mathematical models. As a result, the process of approximation has become a time-honored trade-off that engineers have accepted and should be willing to interrogate its consequence. Viewed through this lens, being aware of the approximate nature of simulation requires analysts to be mindful of errors that arise from simulation and others closely linked with using finite element simulation software as an engine of inquiry to analyze and predict the behavior of physical entities. Nonetheless, we will explore methods of minimizing errors in finite element simulations (through convergence analysis, verification, and validation) in subsequent chapters of this book.

Meanwhile, as a subset of the CAE skills, finite element simulation was once delegated to specialists within engineering firms. However, as the line between engineering analysts and designers blurs with the proliferation of software such as SOLIDWORKS, many engineers are now required to be both familiar and proficient with complex engineering analyses related to the performance evaluations of products. It is hoped that this book serves you in the journey to acquire proficiency in this regard or will, at the very least, point you in this direction.

Applications of FEA

Although FEA gained tremendous traction from its attempts to solve the problems of structural mechanics, today, the successful applications of the methods span numerous subfields of engineering, ranging from flow analysis to thermal, electric, and magnetic fields. A non-exhaustive list of domains of the applications of FEA are presented in *Table 1-1*:

Domain	Examples
Structural analysis	• The analysis of pressure vessels • The static analysis of aircraft structures • The analysis of civil engineering structures (for example, shear walls, bridges, concrete, and frames) and the design of automotive parts for safety • The analysis of load-carrying mechanical structures (such as gears, axles, cranes, and machine frames) • The stress analysis of nuclear reactors • The static analysis of vehicle chassis
Dynamics/ acoustics analysis	• Crash analysis, impact analysis, and modal analysis • Frequency response analysis of structures under time-dependent loads, such as turbine blades and automotive parts • The acoustic design of buildings, and the design of powertrain acoustics
Heat conduction	• Combustion process analysis and the thermal flux of engine compartments • The design of cooling systems (for example, for buildings and electronics)
Biomedical engineering	The modeling of heart valves, the design of new implants, the design of prosthetics, the stress analysis of bones, and more
Flow dynamics	Blood flow analysis in arteries, seepage analysis, the aerodynamics design of cars and aircraft structures, and more

Table 1-1: Domains of applications

Implementations of FEA

Meanwhile, for simple problems, the FEA can be coded in almost any programming language. However, such programs are usually limited in scope and are often less useful for engineers dealing with the performance analysis of complex parts or assemblies. As a consequence, there are many commercial implementations of FEA.

Two categories of FEA-related software have emerged from the implementation by various corporations and entities:

- Analysis-oriented FEA software
- Design-oriented FEA software

The first category encompasses commercial implementations of FEM such as ABAQUS, ADINA, DEFORM, ANSYS, MSC NASTRAN, and COMSOL, among others. Each piece of software in this category predominantly exists as an analysts' tool. They have a comprehensive set of libraries and elements for the advanced analysis of multiphysics engineering systems. However, they tend to have a rather steep learning curve. In contrast, the software in the second category, under which SOLIDWORKS Simulation belongs, is principally developed for **three-dimensional** (**3D**) CAD modeling. However, they offer simulation suites that can be used for various analyses using the FEM. Due to the close integration between the modeling and analysis environments, the latter category generally does the following:

- It facilitates faster learning of the intricacies of FEA.
- It has a familiar and less intimidating interface for beginners.
- It has a relatively shallow learning curve for most engineers that are already familiar with the modeling interface.

Nevertheless, there are elements of overlap in both categories. For instance, a majority of the specialist FEA applications in the second category are also conferred with CAD interface for part modeling. Moreover, all implementations of FEM conceptually follow and require these three phases for product simulations:

- A preprocessing phase
- A solution phase
- A postprocessing phase

The **preprocessing** phase involves *idealization* (which translates to the transformation from a physical world to a computational domain), *model generation* (that is, defining the geometric domain), *mesh generation* (that is, creating elements and nodes), and the *supplying of input data* (for example, material properties, loads, and physical constraints).

In the **solution** phase, the governing algebraic equation in matrix form that maps to the behavior of the computational domain is solved using a numerical method. For this phase to happen, the application software will often require the user to provide details (specifically, sufficient boundary conditions) that ensure the satisfaction of compatibility and equilibrium conditions.

The **postprocessing** phase involves evaluations and interpretations of the computed solutions generated by the simulation and possibly an examination of the correctness. In specific terms, activities that fall under this phase encompass things such as the plotting of results, the retrieving of deformed shapes, the examination of critically-stressed areas within the components, and more.

Now that we have covered the background, applications, and some of the basic steps necessary for general finite element analyses, we will move on to introduce the SOLIDWORKS simulation.

Overview of SOLIDWORKS simulation

This section introduces the SOLIDWORKS Simulation, highlights the basic steps required for most simulations, discusses the type of finite elements provided by SOLIDWORKS Simulation, and covers the SOLIDWORKS Simulation license, its computing requirements, and its limitations.

What is SOLIDWORKS Simulation?

SOLIDWORKS Simulation is the implementation of the FEM in the SOLIDWORKS CAD environment by SOLIDWORKS Corporation (whose parent company, Dassault Systèmes, makes the SOLIDWORKS CAD software). The SOLIDWORKS CAD software has a reputation for being user-friendly, and it is clearly a leader in the 3D design modeling market.

Riding on the wave of popularity of SOLIDWORKS as a design modeling tool, SOLIDWORKS Simulation was developed in the same spirit to provide an easy, one-stop platform for design analyses. In addition to this, SOLIDWORKS Simulation is established on the backbone of fast numerical solvers. It simplifies the workflow for obtaining a detailed solution for stress, thermal, frequency, flow, transient, buckling, pressure vessels, and optimization analyses, among others. Fully embedded within the SOLIDWORKS environment, SOLIDWORKS Simulation helps product designers to do the following:

- Reduce the cost of prototyping by facilitating a virtual testing platform in place of costly early-stage physical tests.

- Shorten the concept-to-product timeframe and the time to market.

- Accelerate the analysis of design iterations.

- Evaluate the optimal design with parametric analyses.

- Analyze complex parts and assemblies with support for different material behavior (such as linear or nonlinear).

- Conduct simulation on subassemblies with support for contact and interaction involving machine elements such as bolts, pins, springs, and bearings.

Basic steps in SOLIDWORKS Simulation

In this section, we will highlight the steps required for the analyses of a single-member component and a multi-member assembly using SOLIDWORKS Simulation. The steps are summarized in *Figure 1.1* and *Figure 1.2*, representing the expansion of the phases in FEA that were briefly mentioned in the *Implementations of FEA* section :

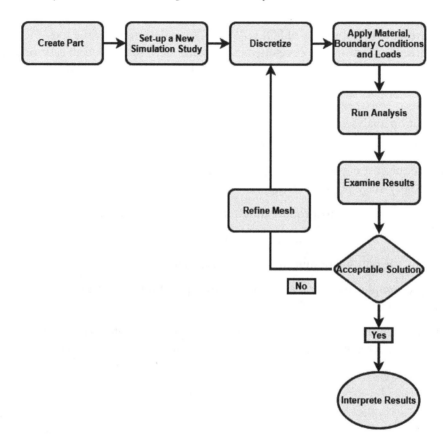

Figure 1.1 – Flowchart for the static analysis of a one-member component

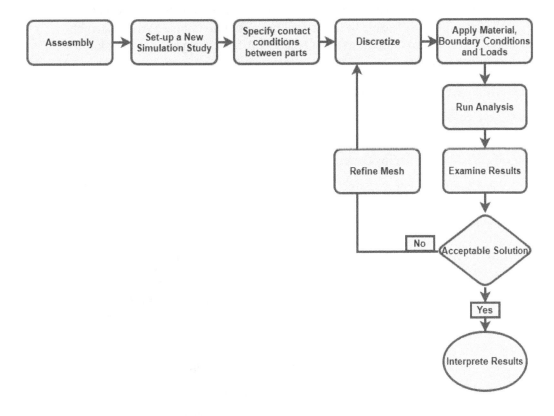

Figure 1.2 – Flowchart for the static analysis of an assembly

A couple of comments regarding the steps indicated in *Figure 1.1* and *Figure 1.2* are provided as follows:

1. The first step to the simulation of a product (such as a part or an assembly) is to create its CAD model. At this stage, all geometric properties are defined. For complicated geometries, the geometry of the structure to be analyzed might have to be defeatured and fine-tuned.

2. Next, the SOLIDWORKS Simulation interface is launched.

3. Discretization of the part or assembly is carried out. Often, discretization is called meshing. This refers to the crucial process of dividing a part or assembly into smaller pieces (similar to LEGO pieces). A few concepts need to be known regarding meshing:

    - Meshing creates *elements* and *nodes*.

    - An element describes a finite-sized division created from the original component to be analyzed.

    - Elements are joined by common points called nodes.

    - Each finite element is characterized by a specific number of degrees of freedom. A degree of freedom is the fundamental field variable calculated during the FEA. For instance, for static analysis problems, the displacement vector is the main degree of freedom during the computation. However, in the case of simulations related to thermal analysis problems, the degree of freedom is temperature.

    - The size and type of elements created during meshing are key to getting accurate results. Typically, the types of elements to be used for analysis become obvious from the nature of the problem. This concept will be developed further throughout the book.

4. After the discretization step, we will specify the following:

    - **Material properties**. For static analysis, this will generally include stiffness information such as Young's modulus and Poisson's ratio.

    - **Loads**. A variety of loads can be applied within the SOLIDWORKS Simulation interface, ranging from axial load, transverse load, torsional load to pressure load.

    - **Fixtures**. In the language of SOLIDWORKS Simulation, the word "fixture" is used to indicate boundary conditions. Meanwhile, boundary conditions generally refer to physical constraints on the movement of specific joints or segments of a load-bearing structure. They arise from the presence of supports used to ensure that a structure being analyzed is properly constrained to prevent rigid body motion during the application of external loads.

    - **Connections**. In the language of SOLIDWORKS Simulation, the connections settings comprise the contact condition that is required anytime two or more components touch each other before or during the simulation process. This might arise from welding, bonding, riveting, or various other types of joining of a practical nature. SOLIDWORKS Simulation provides a variety of contact types that will be explored as we progress in our exploration of the software.

5. Finally, we run the analysis, then obtain and interpret the results.

Information

In FEA, elements of different shapes, **degrees of freedom (DOF)**, and complexity exist. In principle, when the term DOF is used in mechanics, it denotes the number of independent quantities required to describe a displaced or perturbed state of a structure. For static problems that are the focus of this book, *we will be using DOF to refer to the number of possible displacement components at nodes of a specific finite element.* Note that a comprehensive account of the mathematical derivations for a wide variety of elements is not addressed in this book. Such derivations can be found in many of the books on the mathematical foundation of the FEM such as *[3]* and *[4]*.

Elements within SOLIDWORKS Simulation

SOLIDWORKS Simulation has three major families of elements that are used in the performance analysis of components:

- Continuum elements:

  - Solid elements

  - **Two-dimensional (2D)** plane elements

- Structural elements:

  - Beam elements

  - Truss elements

  - Shell elements

- Special elements

While these elements will be rigorously explored in subsequent chapters, *Table 1-2* highlights three representative cases of when to use these elements.

Generally, a **solid** element is used for bulky models with considerable thickness and volume. **2D plane** elements are employed for the 2D analysis of members (such as axisymmetric, plane stress, or plane strain problems). **Beam** and **truss** elements are used for the analysis of structural members that have one of their dimensions far greater than the dimensions of their cross-sections. Shell elements are deployed for thin-walled members. The special elements mostly connect elements such as springs elastic supports, and more:

Element family	Structure	Meshed equivalent
Tetrahedral solid element	Implant	
Triangular shell element	Vase	
Beam element	Gantry frame	

Table 1-2: Discretization and the major types of elements

Types of SOLIDWORKS Simulation license

SOLIDWORKS Corporation offers three types of license for SOLIDWORKS Simulation:

- SOLIDWORKS Simulation Standard
- SOLIDWORKS Simulation Professional
- SOLIDWORKS Simulation Premium

Of these three, the premium license is the most comprehensive in terms of capability. The professional license does not support nonlinear and composite analyses. The standard license is even more limited in terms of the scope of analyses it supports. For this book, the premium license is employed.

Information

To read more about the kinds of analyses that can be carried out with each of the previously mentioned licenses, please visit `https://www.solidworks.com/product/solidworks-simulation`.

Computing requirements

SOLIDWORKS is a memory-hungry application. This is understandable given the functionalities that are packed into this amazing piece of software. For best performance, the recommendation listed in *Table 1-3* is suggested for PCs or laptops to be used for basic analysis with the SOLIDWORKS Simulation:

Components	Recommendation
Processor	3–4 GHz
Operating system	Windows 10, 64-bit
Graphics card	Mid-range to high-end graphics cards such as, but not limited to, NVIDIA Quadro P, GP, GV series, and AMD Radeon Pro WX
RAM	8 GB (minimum)

Table 1-3: System requirements

Information

For further information about system requirements, beyond the details in *Table 1-3*, head over to `https://www.solidworks.com/support/system-requirements`.

What are the limitations of SOLIDWORKS Simulation?

While SOLIDWORKS Simulation is a powerful tool that can be used for numerous kinds of analyses of products and components, it is worth mentioning that it has a limited number of elements in its library. This point should be borne in mind while dealing with multiphysics problems for which a suitable element for the analysis might not exist in the SOLIDWORKS Simulation library. Besides, you should always find alternative methods to determine the accuracy of the results retrieved from the SOLIDWORKS Simulation. This is known as validation, and it can be done via experiments or analytical techniques at one stage of the product development phases. The approach to such methods of validation through experimental stress techniques is not covered in this book (a classic reference is a text by *Dally* and *Riley [5]*).

This wraps up our presentation of the overview of the SOLIDWORKS Simulation. In the next section, we will do a cursory examination of the SOLIDWORKS interfaces.

Understanding the SOLIDWORKS interfaces

The main focus of this section is to briefly introduce the SOLIDWORKS interfaces. Since we are going to be interacting with the interface in the rest of the book, only a few of the features are examined. Nonetheless, it is worth pointing out that the SOLIDWORKS Simulation interface is closely linked with the SOLIDWORKS modeling environment, and both require that you have the correct license. With SOLIDWORKS installed on your PC or laptop, the interfaces are accessed by following the steps laid out in the subsections that follow.

Getting started with the SOLIDWORKS modeling environment

This subsection illustrates a brief interaction with the SOLIDWORKS 2021-2022 modeling environment. The steps focus on the use of a single-part component to reveal the simulation environment. Let's start by launching the SOLIDWORKS application and then navigating to the modeling environment by following these steps:

1. Choose **File** from the main menu.
2. Click on **New**.
3. Select **Part** and click on **OK**:

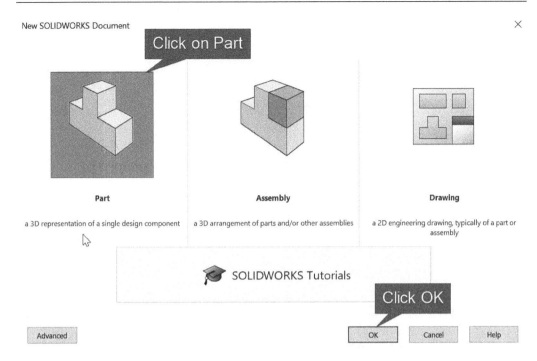

Figure 1.3 – The steps for launching the modeling environment

The modeling environment is launched after completing the preceding steps, as shown next. Generally, the modeling environment features many items, as shown in *Figure 1.4*. This includes the following:

- **Menu Bar**: This provides access to different kinds of commands that the software offers.

- **Command Manager Tab**: This encompasses a series of tabs segregating the commands for many specialized tasks.

- **Feature Manager Tree**: This acts as a record of features that are created in the graphics window, often representing these features in the order in which they are created.

- **Document Window**: This is used to navigate between different windows in the graphics area.

- **Graphics area**: The major area for modeling and simulation activities:

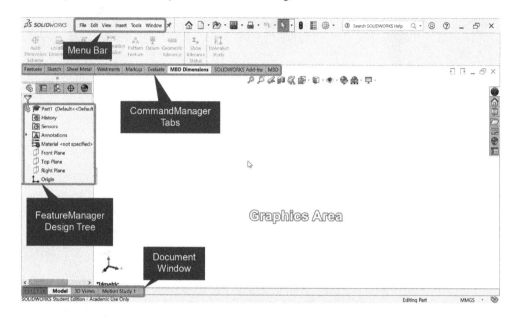

Figure 1.4 – The SOLIDWORKS 2021-2022 user interface

With the basic information about the interface detailed, let's now take a brief look at how to activate the simulation environment. We will come back to this activity in subsequent chapters in more detail.

Activating the SOLIDWORKS Simulation environment

Launch the SOLIDWORKS Simulation interface by following these steps:

1. Activate the simulation add-in by performing the following:

    I.   Clicking on the **SOLIDWORKS Add-Ins** tab (please refer to *Figure 1.5*).

    II.   Selecting **SOLIDWORKS Simulation**:

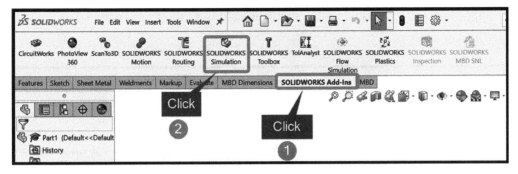

Figure 1.5 – Activating the SOLIDWORKS simulation add-ins

The **Simulation** tab becomes active, as shown next. However, notice that when the SOLIDWORKS Simulation becomes activated, most of the icons are gray, as shown in *Figure 1.6*. This arises from the fact that no analysis has been defined yet.

2. Start a new analysis by performing the following:

    I.   Open a CAD model of a part or a component to be analyzed.

    II.   Click on **New Study**, as shown in *Figure 1.6*:

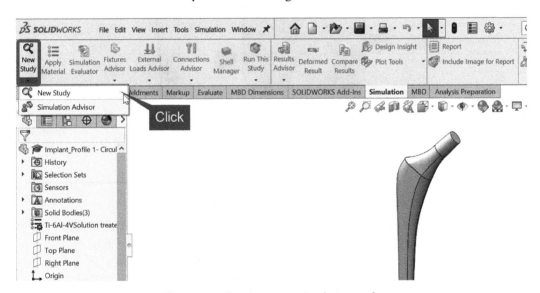

Figure 1.6 – Starting a new simulation study

3. Select a **Study** type (in this book, we will be restricted to static analysis).

4.   Supply a descriptive name for the simulation study, as shown in *Figure 1.7*.

5.   Click on **OK**:

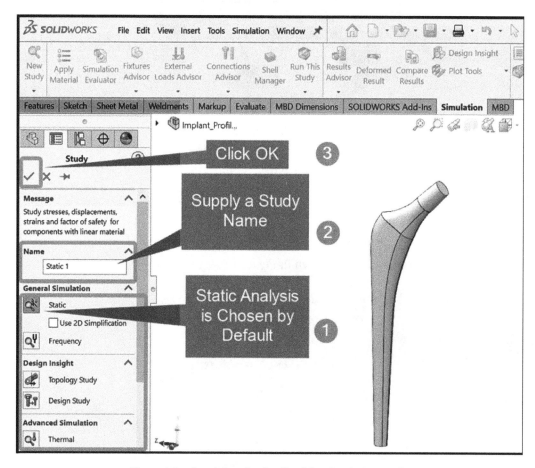

Figure 1.7 – Supplying the details of the simulation study

In response to the preceding steps, the simulation study environment is activated, as indicated in *Figure 1.8*. There are a few things to pay attention to in this screenshot. For one, the different icons that were previously gray underneath the Simulation tab in *Figure 1.6* are now active in *Figure 1.8*. Further, the simulation tree manager and the study tab (at the base of the screen) have both appeared:

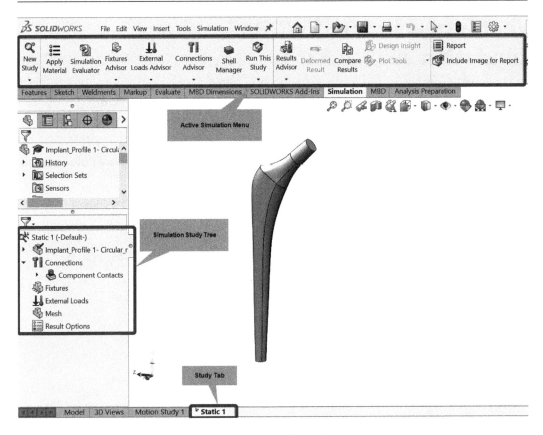

Figure 1.8 – Simulation study tree

The SOLIDWORKS Simulation environment is better explored within the context of simulation problems. Accordingly, rather than detailing all the features here, we will further examine them comprehensively in subsequent chapters and reveal the power of this simulation engine for the analysis of various types of problems.

In the next section, we will briefly highlight some of the important updates in SOLIDWORKS Simulation 2021-2022.

What is new in SOLIDWORKS Simulation 2021-2022?

SOLIDWORKS 2021-2022, upon which this book is based, is the latest version of SOLIDWORKS with significant improvement in functionality and performance. In terms of its look, SOLIDWORKS 2021-2022 appears similar to the previous version of SOLIDWORKS (specifically, the 2020-2021 version). However, there are important differences across many phases of the software. Nonetheless, when it comes to the 2021-2022 version of SOLIDWORKS Simulation, a few of the updates are highlighted here:

- *Change in terminology for the items under the Connections folder within the simulation study tree*: For instance, if we import the model of an assembly into the modeling environment and then launch a **New Study** (as done in *Figure 1.6*), the look of the Simulation study tree in the 2021-2022 and the 2020-2021 versions will be similar to *Figure 1.9*:

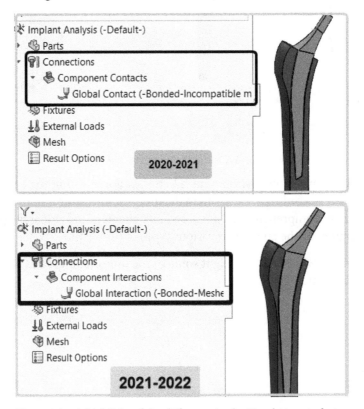

Figure 1.9 – A highlight of the difference in the Simulation study tree

As you can see, **Component Contacts** is now known as **Component Interactions**, while **Global Contact** becomes **Global Interaction**.

- *Update to the Static Options dialog box*: After launching a new study environment, you can examine the Static Options dialog box by following *Figure 1.10*:

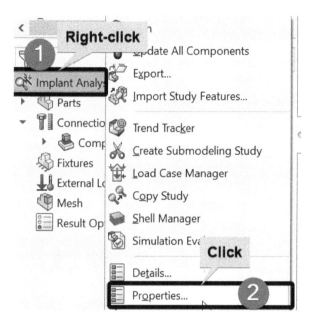

Figure 1.10 – Initiating the static options dialog box

After clicking on **Properties...**, the Static options dialog box appears. As shown in *Figure 1.11*, the **Static** options dialog box for the 2021-2022 version has a more streamlined interface for modifying various study properties.

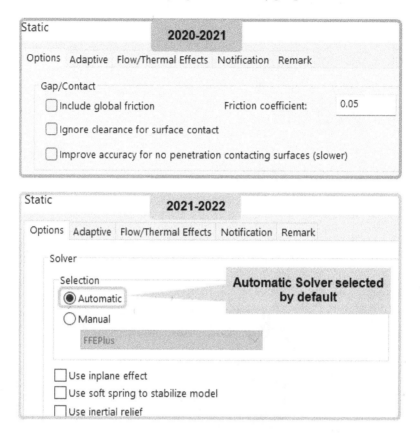

Figure 1.11 – Partial views of the static options dialog box

Additionally, as you will note from *Figure 1.11*, in the 2021-2022 version, the **Automatic Solver** is selected by default within the static options dialogue box. And talking about the static options dialog box, the number of solution Solvers available in the 2021-2022 version is the same as the earlier version, as shown in *Figure 1.12*. However, the **FFEPlus** solver, which is based on an iterative technique is now more powerful (this is true for the other solvers as well):

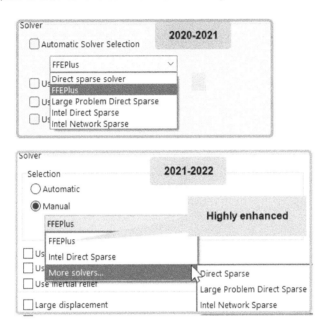

Figure 1.12 – The Solver options within the static options dialog box

Apart from the aforementioned update, we can now shift our attention briefly to the update to the Connections folder's sub-items.

- *Update to the Connections folder context menu.* It is shown in *Figure 1.9* that there is a change in terminology concerning the **Connections** folder sub-items under the Simulation study tree. The update is deeper than what was highlighted in *Figure 1.9*. To see another update, you should right-click on the **Connections** folder. From the right-click context menu, you will notice that the items named **Contact Set...** and **Component Contact...** are now referred to as **Local Interaction...** and **Component Interaction...** , as depicted in *Figure 1.13*:

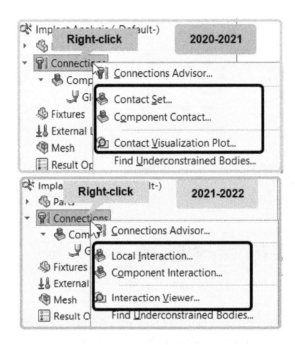

Figure 1.13 – Highlight of the update to the Connections folder context menu

We will expand on this change in more detail in *Chapter 6, Analyses of Components with Solid Elements*, and *Chapter 7, Analyses of Components with Mixed Elements*.

- *Update to the Mesh PropertyManager*: If you right-click on the **Mesh** folder within the Simulation study tree and then select **Create Mesh**, you will observe a difference in the arrangement of the meshing engines, as shown in *Figure 1.14*:

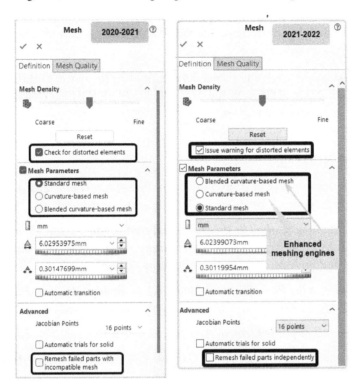

Figure 1.14 – Highlight of the change for the Mesh PropertyManager

While the names of the meshing engines remain the same, as shown in *Figure 1.14*, the **Curvature-based mesh** and the **Blended curvature-based** meshing engines have undergone serious updates to facilitate enhanced accuracy of the simulation results. Again, we will revisit the issues around meshing in the second and third sections of the book.

This ends our discussion of a few of the differences that exist in the 2021-2022 SOLIDWORKS Simulation. So far, we have primarily focused on the updates that will be discussed in the later chapters of the book. For a more detailed look at the significant enhancements across all aspects of SOLIDWORKS, in general, and SOLIDWORKS Simulation, in particular, you should check out `https://www.solidworks.com/product/whats-new`.

Summary

This chapter provided a short overview of the importance, applications, and basic concepts of finite element simulation (such as discretization, elements, the types of elements, nodes, the main phases in finite element simulation, and more). We also initiated our exploration of the theme of this book by introducing the SOLIDWORKS general interface and the SOLIDWORKS Simulation interface.

Subsequent chapters of the book will take a detailed look at the use of SOLIDWORKS Simulation for the analyses of different kinds of structures. In the next chapter, we will examine the analysis of bars and trusses.

Further reading

- [1] *The finite element in plan stress analysis, in Proceedings of the 2nd ASCE Conference on Electronic Computation, R. W. Clough, Pittsburgh, PA, 1960,*

- [2] *The finite element method in structural and continuum mechanics: numerical solution of problems in structural and continuum mechanics, O. C. Zienkiewicz and Y. K. Cheung, London; New York: McGraw-Hill (in English), 1967.*

- [3] *Introduction to the Finite Element Method, J. N. Reddy, McGraw-Hill Education, 2019.*

- [4] *Fundamentals of Finite Element Analysis, D. V. Hutton, McGraw-Hill, 2003.*

- [5] *Experimental Stress Analysis, J. W. Dally and W. F. Riley, McGraw-Hill, 1978.*

2
Analyses of Bars and Trusses

This chapter demonstrates the SOLIDWORKS simulation procedure for structures that primarily support axial loads. The simplest form of this type of structure is known as bars or rods. In the more complex form, they are known as plane and space trusses (which are just the two-dimensional and three-dimensional arrangements of bars, respectively). By the end of this chapter, you will be familiar with the procedure for the simulation of the aforementioned structures. Against this backdrop, the focus of this chapter is anchored on the following topics:

- Overview of static analysis of trusses
- Strategies for the analysis of trusses
- Getting started with truss analysis via SOLIDWORKS Simulation

Technical requirement

You will need to have access to the SOLIDWORKS software with a SOLIDWORKS Simulation license.

You can find the sample files of the models required for this chapter here: `https://github.com/PacktPublishing/Practical-Finite-Element-Simulations-with-SOLIDWORKS-2022/tree/main/Chapter02`

Overview of static analysis of trusses

This section provides basic background information about bars and trusses. It highlights the objectives of analyzing these structures and their applications.

Let's start with some basic definitions. A bar is a structure that is designed to support simple forces along its axis (such as tensile and compressive loads). On the other hand, a truss represents a collection of bars that are arranged as one or more units of triangulated frameworks. Discussions on the analysis of either of these types of structure (that is, a bar or a truss) often deserve separate standalone chapters. Nonetheless, since the analysis of bars is simpler than that of trusses, we shall allocate more time to the simulation of trusses with the understanding that the same knowledge carries over to the analysis of simple bars.

> **Note**
> Within the subject of mechanics, a structure broadly refers to a body or a collection of bodies designed to carry loads. Most structures are **three-dimensional (3D)** in nature. But for ease of analysis, engineers often leverage approximations that facilitate the use of **one-dimensional (1D)** members (such as a bar, a shaft, a beam, a column, and so on) or **two-dimensional (2D)** approximations (such as plates and shells, and so on) to reduce the computational burden of complex 3D analyses.

You will have seen the application of truss structures in various forms around you. Some of these are shown in *Figure 2.1*. Typically, truss structures are featured prominently in the design of cranes, truss booms, telecommunication towers, masts, electric pylons, roofs, bridges, and so on.

From an engineering performance analysis point of view, we conduct static analyses of bars and trusses with the following objectives:

- Determining the internal forces and consequently the stresses that developed in the members
- Evaluating the axial deformation, the members experienced upon loading

In bars and trusses, the axial deformations manifest in the form of shortening (contraction) or lengthening (extension) of a member's length. Consequently, a combination of compressive and tensile normal strains/stresses develops in these structures. Together, the stress and the deformation data that we retrieve from simulations help in determining the right geometric sizing for the members (called proportioning). But crucially, the results contribute towards our ability to design these structures to hedge against unwanted failure or excessive deformation during in-service usage. For brevity, in the rest of this chapter, we shall be using the term *truss* as a shorthand for axially loaded structures.

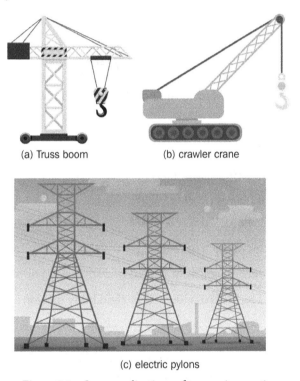

(a) Truss boom   (b) crawler crane

(c) electric pylons

Figure 2.1 – Some applications of trusses in practice

Before we get deep into analysis, it is important to be aware of the following technical points that are frequently considered for the computer analysis of trusses:

- The members of a truss are straight and have uniform cross-sections.

- The ends of an individual member of the truss are connected to the ends of other members at joints via frictionless pins. In practical scenarios, such joints may be formed by rivets/bolts/ball and socket joints or via welding to a gusset plate.

- Forces and supports are applied only at the joints of a truss.

With the background information provided about trusses, let's now take a look at strategies for their simulations in the next section.

Strategies for the analysis of trusses

This section describes the structural details and the modeling strategies for the simulation of truss structures. It also highlights the major features of the **truss element** within the SOLIDWORKS Simulation library.

Structural details

Trusses can be designed to function under a wide spectrum of load-supporting applications. However, irrespective of what form they take in appearance, a consistent set of parameters is employed for their analyses. This implies that irrespective of the form of the truss you are analyzing, you will need to know the following technical information before venturing into the analysis:

- The dimensions of the truss:

    A.   The details of the cross-section

    B.   The geometric length of each member

    C.   The orientation angles of the members

- The material properties of the truss members
- The loads applied to specific joints of the truss
- The support provided to the truss to prevent rigid body motion

> **Note**
>
> In practice, a chunk of time will be spent spelling out the scope of the problem relating to structures'/products' design and analyses. Things such as what the magnitude of the load should be. In what environment will the truss be used? What materials should be used? What types of support are required? What dimensions should be assigned to each member? And so on. In short, to achieve meaningful simulation results, attention should be paid to project specifications and parameters before commencing the simulation tasks. That said, for much of this book, these details will be provided so we can focus squarely on the simulation tasks without getting bogged down by the time-consuming iterative conceptual design tasks.

Modeling strategy

When we analyze truss structures via the finite element simulation method, a basic strategy is to take a structure such as that in *Figure 2.2 (a)*, split it into its constituent members (*Figure 2.2 (b)*), and then treat each member as a truss element. So, in the end, a whole structure formed from the assembly of different members is represented by the collective behavior of individual truss elements.

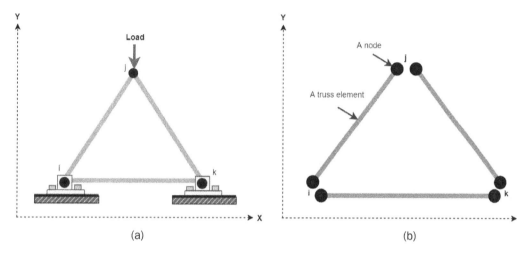

(a)   (b)

Figure 2.2 – (a) Illustration of a simple truss structure; (b) splitting the structure into elements

Figure 2.3 illustrates the other aspect of the strategy for the simulation of trusses:

1. The first step is modeling the skeletal arrangement of the truss or a collection of bars within the SOLIDWORKS model environment.

2. This is then followed by the conversion of the skeletal model into a weldment profile, transforming the weldment model into a finite element model.

3. Finally, run the analysis to obtain the results (within the SOLIDWORKS Simulation window).

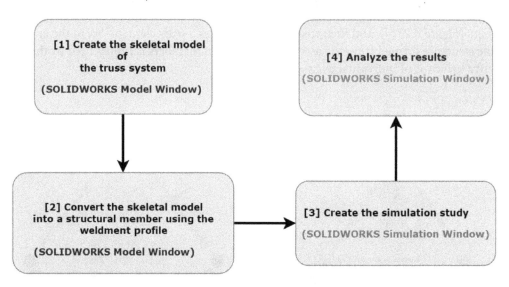

Figure 2.3 – Major steps in the static analysis of trusses

Characteristics of the truss element in the SOLIDWORKS Simulation library

In the traditional theoretical treatment of the finite element simulation method, the **truss element** is generally known to be a plane element with two nodes and two degrees of freedom. Within the SOLIDWORKS Simulation environment, the **truss element** is a 3D two-node element with three degrees of freedom per node (that is, translational displacements about the x, y, and z axes). The beauty of the 3D **truss element** in SOLIDWORKS is that it can be used to analyze 1D bars, 2D plane trusses, and 3D space trusses.

We will go through a case study in the next section to explore further details about the simulation of a loaded truss structure.

Getting started with truss analysis via SOLIDWORKS Simulation

In this section, we will demonstrate the use of the **truss element** with a practical case study. Primarily, we will analyze the structural performance of a crane used in the spatial positioning of heavy objects on a mega building construction site. The example is inspired by exercise 4.11 in the textbook by *Megson [1]* (see the *Further reading* section). It is a practical problem that can only be solved satisfactorily via computer analysis.

Time for action – Conducting static analysis on a crane

Problem statement

Consider the 2D representation of a crane shown in *Figure 2.4*, which is to be analyzed based on the placement of *1500 kN* and *2000kN* weights at points R and W, respectively.

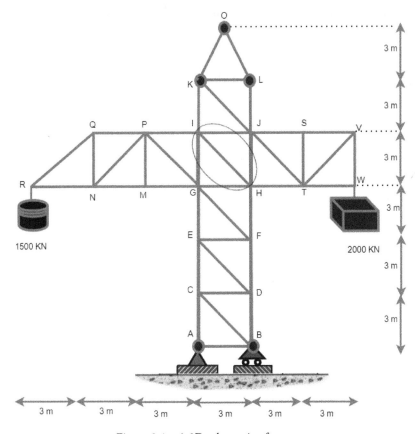

Figure 2.4 – A 2D schematic of a crane

For convenience, we shall consider that the members of the crane are derived from tubular alloy steel. For this material, the Young's modulus, E, is taken as *210 GPa*. The members have the same cross-sectional detail characterized by external and internal diameters *200 mm* and *80 mm*, respectively. Using SOLIDWORKS simulation, we want to answer the following questions:

- What is the maximum resultant deformation of the truss upon the application of the loads?

- What is the distribution of the factor of safety of the members of the crane upon loading?

- What is the internal force/stress that developed in the member IH?

Part A – Creating the sketch of the geometric model

The first step in any analysis is to have a model of the structure to be analyzed. This section centers around the creation of the basic geometric lines describing the structure.

Starting up the SOLIDWORKS interface

To create the skeletal sketch of the model, launch SOLIDWORKS and follow these steps:

1. Click **File**.

2. Select **New**.

   After *step 2*, the **New SOLIDWORKS Document** screen appears. We are interested in creating the model using the **Part** modeling environment.

3. Select **Part** and then click **OK** (*Figure 2.5*).

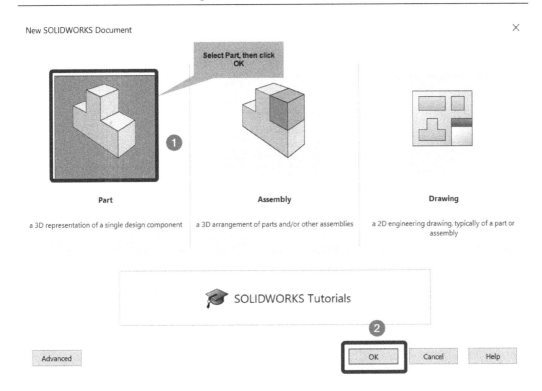

Figure 2.5 – Creating a new Part document

The SOLIDWORKS user interface opens up and we can start sketching right away. But before that, it is a good idea to ensure you are using the right unit of measurement.

Setting the dimensions

Look at the lower-right corner of the **graphical user interface (GUI)** to ensure that the current unit of measurement is the **MMGS (millimeter, gram, and second)** system of units, as shown in *Figure 2.6*.

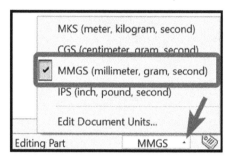

Figure 2.6 – Setting the unit of the document

Once it is confirmed that the right unit is set up, you may save the file as Crane.

Sketching the lines describing the geometry of the crane

The view of the crane shown with the problem statement represents the front view. Consequently, we shall sketch the geometric model of the truss on the front plane:

1. Click on the **Sketch** tab.
2. Click the **Sketch** manager, as shown in the following figure.
3. Choose **Front Plane**.

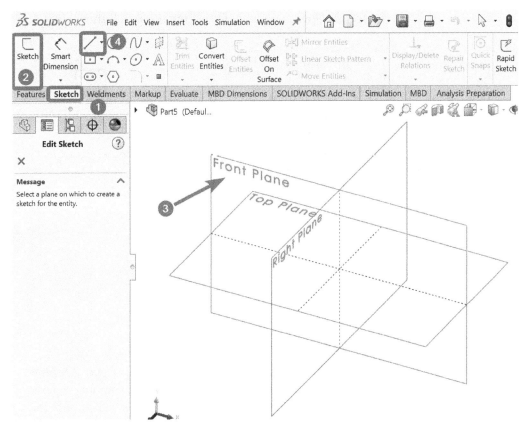

Figure 2.7 – Commencing the sketching task

4. Create the sketch of the crane with the line sketching tool. Ensure the sketch is symmetric around the origin (coordinate 0 of the graphics window), as shown in *Figure 2.8*.

5. Save and exit from the sketch. Note the symbol for exiting the sketch shown in
    Figure 2.8.

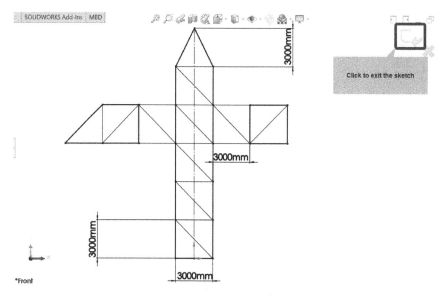

Figure 2.8 – A line-based geometric model of the crane

After completing the sketch, it should look as in the screenshot shown in *Figure 2.8.* Now,
the sketch we have created in this section is just a series of lines with no volume property.
As such, it cannot be used for structural analyses. In the next section, we will convert the
sketched line-based model into a structural model with a volume property.

Part B – Converting the skeletal model into a structural profile

This section explains how to convert the line sketches we created in the preceding
section into a structural model using a special functionality in SOLIDWORKS called the
weldments tool.

Introducing the weldments tool

The **weldments** tool in SOLIDWORKS makes it easy to prescribe the cross-sectional details for sketched lines. In other words, it facilitates the transformation of lines with no volume properties into structural members with volume properties that are suitable for realistic engineering simulation.

The weldments tool can be used with both 2D and 3D sketches. What is more interesting about this tool is that it provides us with access to a handful of relevant structural profiles, such as those listed in *Table 2.1*.

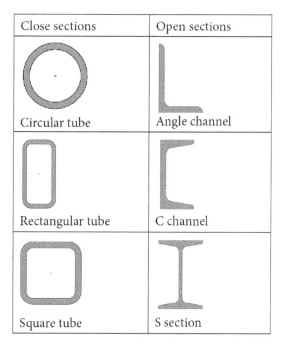

Close sections	Open sections
Circular tube	Angle channel
Rectangular tube	C channel
Square tube	S section

Table 2.1 – Samples of in-built structural cross-sections in the weldment library

The profiles highlighted in *Table 2.1* are stored in the SOLIDWORKS installation folder. For instance, for laptops/PCs with a Windows operating system, the folder is located at `Drive:\Program Files\SOLIDWORKS Corp\SOLIDWORKS\lang\english\ weldment profiles.`

Note that `Drive` is a placeholder for the storage drive containing the SOLIDWORKS installation folder on your device. Scrutinizing the aforementioned directory address, you will notice that the profiles are contained in the parent folder named `weldment profiles`. This folder contains sub-directories, as shown in *Figure 2.9 (a)*. Further, opening any of the sub-directories will expose relevant files with the `.sldlfp` extension, as shown in *Figure 2.9 (b)*.

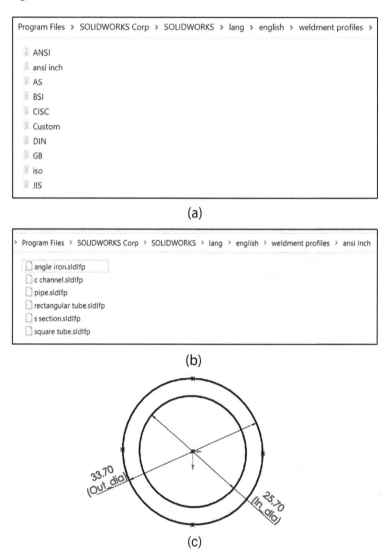

(a)

(b)

(c)

Figure 2.9 – The content of the weldment profiles directory: (a) the two original sub-folders; (b) the library's six profile files in one of the original folders; (c) the dimensions of the largest ISO tube profile in the weldment library

Now, the profiles that are bundled with the **weldments** tool have pre-defined dimensions, but these dimensions may differ from what we need. Consequently, it is common to have to adjust or edit these profiles to fit our needs. For instance, the external and internal diameters of the cross-section we need are *200 mm* and *80 mm*, respectively (from the problem statement). However, the largest tube within the weldment profiles folder differs from these values, as indicated in *Figure 2.9(c)*. In the next sub-section, we will explore how to edit the profile for our needs, but we first need to activate the weldment profile.

> **Information**
>
> Our main interest is in using the weldment tool to supply the cross-sectional details of the members. However, it has many features that can be further explored. Relevant details can be found by following this SOLIDWORKS help link: `https://help.solidworks.com/2022/english/SolidWorks/sldworks/c_Weldments_Overview.htm?verRedirect=1`.

Activating the weldments tool

Check the set of items in your SOLIDWORKS's **CommandManager** tab to see if the **Weldments** tab is present. If the **CommandManager** tab is missing the **Weldments** tab, then follow these steps (summarized in *Figure 2.10*):

1. Right-click on the **CommandManager** tab to bring up the option to show more tabs.

2. Move your cursor over the word **Tabs**.

3.   Click on **Weldments** (this will activate the **Weldments** tab).

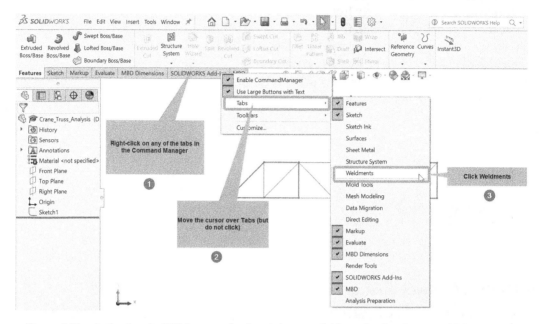

Figure 2.10 – Activating the Weldments tab when it is unavailable under the Command Manager tab

With the **Weldments** tab on, the next sub-section illustrates how to convert the sketched lines to structural members.

Adding structural properties

Follow these steps to transform the sketched lines into structural members:

1.   Click on the **Weldments** tab (*Figure 2.11*).

2.   Click on **Structural Member**.

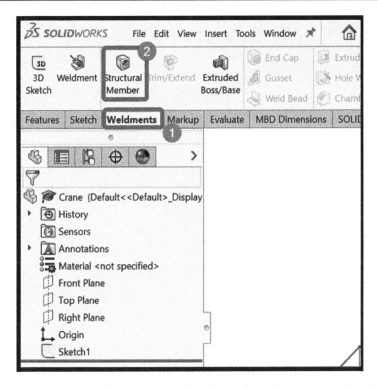

Figure 2.11 – Initiating the conversion of the sketched lines into structural members

Once you have clicked on the **Structural Member** command, the **Structural Member** property manager window will open on the left side of the GUI. Select the options highlighted in *Figure 2.12*.

3. Under **Standard**, choose **iso**.

4. Under **Type**, select **pipe**.

5. Under **Size**, select **33.7 x 4** – this refers to a tube with an external diameter of 33.7 mm and a thickness of 4 mm.

6. Click **New Group** to form **Group 1** and select the lines shown in *Figure 2.13a*. After completing the selections for **Group 1**, proceed to form **Group 2** by selecting the lines highlighted in *Figure 2.13b*.

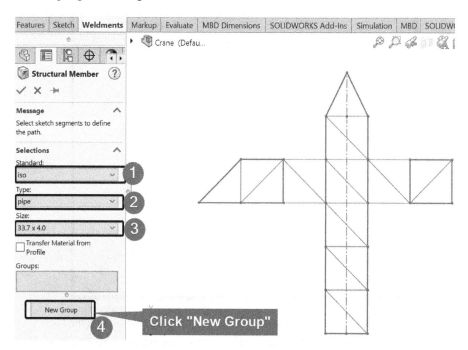

Figure 2.12 – Specifying options for the profile of the cross-section

For ease of forming the structural members, *Figure 2.13 (a-f)* illustrates the series of lines to be selected for each group.

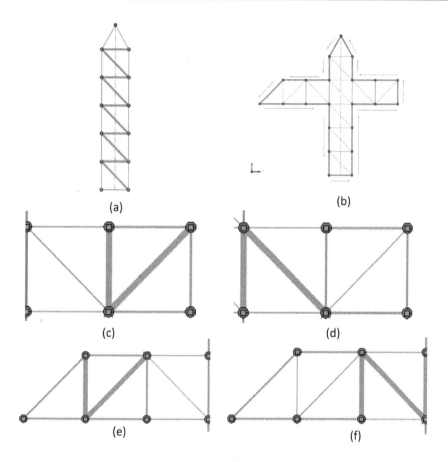

Figure 2.13 – Path segments for the weldment profiles

7. Form all the six groups (as indicated in *Figure 2.14*).

8. Under the settings options, ensure that the **Apply corner treatment** box is unticked.

9.   Click **OK**.

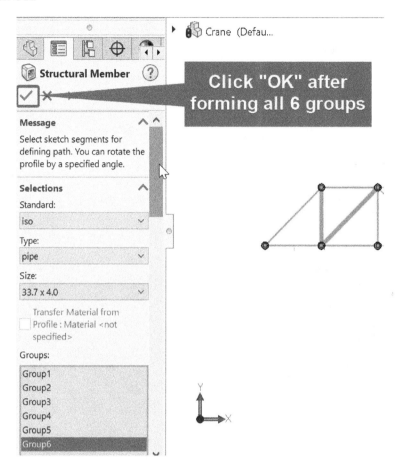

Figure 2.14 – Finalizing the selection for the cross-section

By completing *steps 1-9*, the **Feature Manager** tree will appear with some additional items. Five of these are highlighted in *Figure 2.15*:

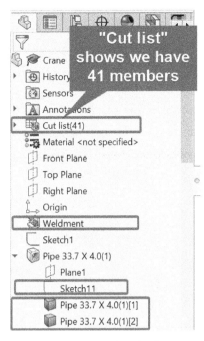

Figure 2.15 – New items within the feature manager

- **Cut list (41)**: In general, this is a folder that contains the details of the weldment items in the current part file. As you can see, it shows that we have a total of 41 weldment items that make up our structure.

- The **Weldment** symbol. This appears once you click on **Structural Member** under the **Weldments** tab.

- *The Structural Member symbol*: This also appears in response to using the **Structural Member** command.

- **Sketch11**: This is the sketch of the cross-section of our weldment profile.

- **Pipe**: This is the main branch of the collection of extruded bodies representing the 41 structural parts of the crane.

Editing the cross-section for our needs

The cross-section that we have employed from the weldment library is not the same as that stated in our problem statement. Thus, it is necessary to change the dimension of this cross-section to suit our needs. To do this, take the following steps:

1. Right-click on **Sketch11** and select **Normal To** (as shown in *Figure 2.16*).
2. Right-click again on **Sketch11**, but this time, select **Edit Sketch**.

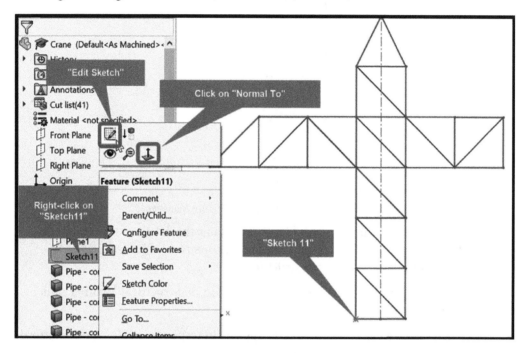

Figure 2.16 – Steps to edit Sketch11

3. Zoom out so you can see the cross-section, as shown in *Figure 2.17a*.
4. Edit the sketch to comply with our desired dimensions, as shown in *Figure 2.17b*.

5. Save and exit the sketching mode, then change the view of the model back to **Front View**.

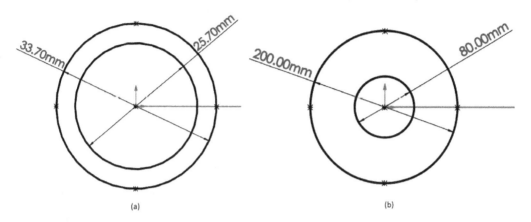

(a)   (b)

Figure 2.17: (a) The original cross-section of the tube; (b) the updated cross-section detail for our analysis

Completing *steps 1-5* wraps up the creation of the structural model with a volume property, and we shall next transition to the initiation of the simulation study.

Part C – Creating the Simulation study

In this section, we will activate the Simulation add-ins, specify the material for the members, indicate how to select the truss element, apply fixtures/loads, and finally, initiate the meshing process.

Activating the Simulation tab and creating a new study

Follow these steps to activate the simulation tab and create a new study:

1. Click on **SOLIDWORKS Add-Ins**.

2. Click on **SOLIDWORKS Simulation** to activate the **Simulation** tab.

Figure 2.18 – Activating the SOLIDWORKS add-ins

3.   With the **Simulation** tab active, click on **New Study**.

Figure 2.19 – Creating a new study

4.   The preceding step launches the **Study** property manager (*Figure 2.20*).

5.   Keep the **Static** analysis option selected by default.

6.   Input a study name within the **Name** box (for example, `Crane Analysis`).

7.   Click **OK** (this closes the **Study** property manager panel and launches the Simulation tree).

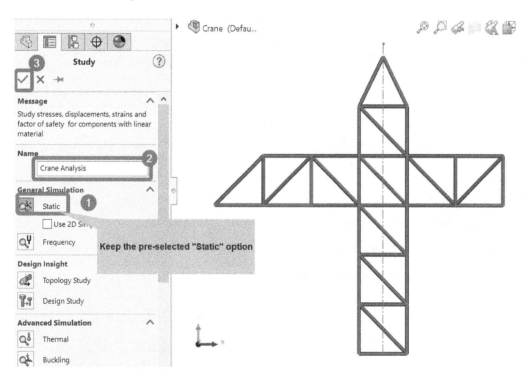

Figure 2.20 – Static study property manager

After completing *steps 1-7*, you will notice the changes shown in *Figure 2.21*. Basically, joints will be imposed at the connection points between the members of the truss. At the same time, the Simulation commands will become available:

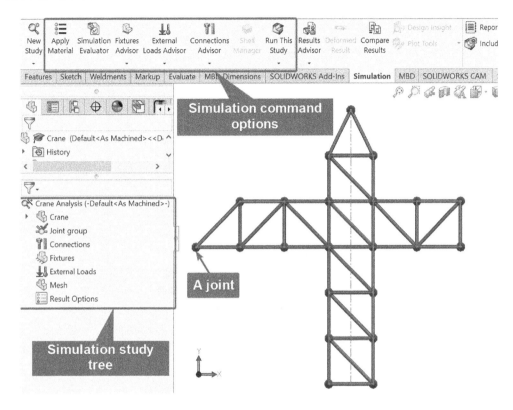

Figure 2.21 – Appearance of joints in the model with the Simulation study tree

In the next sub-section, we will specify the material property for the members.

Adding a material property

Every single member of the crane is assumed to be made of the same material. This makes it easy to apply the material to the members at once:

1. Right-click on the part name – **Crane** (*Figure 2.22*).

2. Click **Apply Material to All Bodies**.

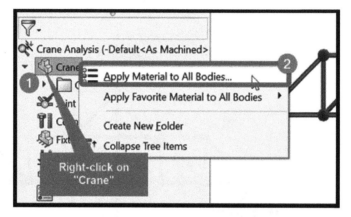

Figure 2.22 – Activating the material database

Step 2 launches the material database, which is shown in *Figure 2.23*. For our analysis, the material that we want is alloy steel, which will be located in the sub-folder called `Steel`. If necessary, expand the `Steel` folder, then perform the following steps.

3. Select `Alloy Steel`.

4. Click **Apply**.

5. Click **Close**.

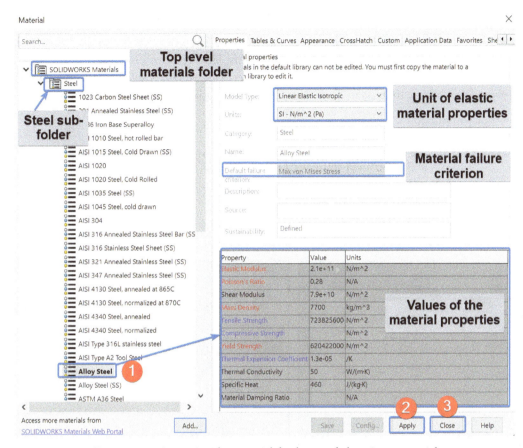

Figure 2.23 – Activating the material database and choosing a material

Before moving to the next sub-section, there are a few features to be observed with the material database. First, the material database is a multilevel directory. At the top layer is **SOLIDWORKS Materials**, then we have sub-folders that contain the same family of materials, and another sub-folder for custom materials. Second, the material property names in *Figure 2.23* are either in black, blue, or red font. In general, the material property names in red font are the ones that are necessary for static analyses. Without values provided for the property names in red font, the simulation will not run. Thirdly, a material failure criterion (**Max von Mises Stress**) and **Linear Elastic Isotropic** material model are pre-defined for the selected material. Lastly, after closing the material database window, a green tick mark (**ü**) will appear on the study name.

Changing from a beam element to a truss element

By default, SOLIDWORKS Simulation treats a structural member that is created using the weldment tool as a **beam element** during the analysis. However, for our case study, what we need is a **truss element**. Therefore, in this sub-section, we will convert all the structural members from beams to trusses. To do this, in the simulation tree, do the following:

1. Expand the folder named `Cut list` – this reveals a set of sub-folders as shown in *Figure 2.24 (a)*.

2. Expand the first sub-folder named `PIPE 200 X 4.0`.

3. Select all members. Note that this is a long list, *Figure 2.24 (b)* shows only a partial view.

4. Click **Edit Definition**.

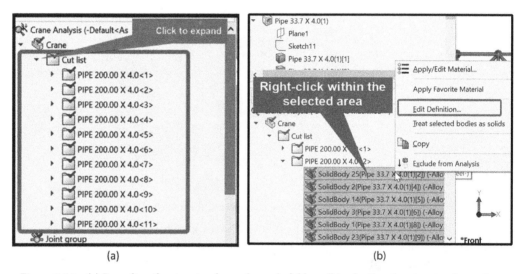

(a)   (b)

Figure 2.24 – (a) Revealing the structural member sub-folders; (b) selecting the structural members

5.   In the **Apply/Edit Beam** property manager that appears, select **Truss** (as shown in *Figure 2.25*).

6.   Click **OK**.

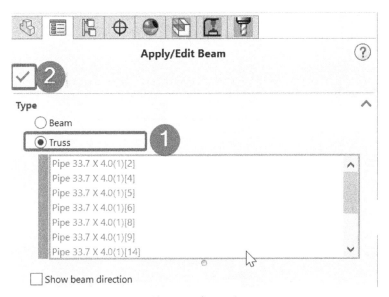

Figure 2.25 – Changing from a beam to a truss

Repeat sub-steps 2-6 for every sub-folders under the **Cut list** to convert the beams to trusses. After completing the conversion of all members, we can now move on to the next phase, which is about the application of constraints or what we often refer to as boundary conditions.

Applying a fixture

A fixture is a constraint that we apply to structures to restrict the movement of its joint/segment when loads are applied. For this analysis, we will apply three sets of restraints to the structural model of the crane:

- A fixture that will prevent the normal movement to the front view of all the joints (that is the vertices of the crane). This needs to be done because we are running a planar (2D) analysis of the crane. However, if we are running a 3D analysis, this is not necessary.

- A fixture that prevents movements in the horizontal and vertical directions at joint *A* (because joint *A* has a fixed support).

- A fixture that prevents the movement along the vertical direction at joint *B* (because this joint is supported by a roller joint).

Let's start with the application of the first fixture to all the nodes by following the steps highlighted next.

Applying a fixture to restrain the Z motion of all nodes:

1. Right-click on **Fixtures**.
2. Pick **Fixed Geometry** from the context menu that appears (*Figure 2.26*).

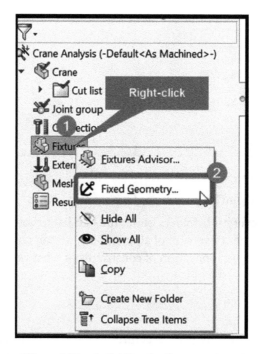

Figure 2.26 – Activating the fixture options

3. Under the **Fixture** property manager that appears, click **Use Reference Geometry** (*Figure 2.27*).
4. Move the cursor into the graphic window and pick all the joints one by one.
5. Click inside the box for the reference plane (labeled 3 in *Figure 2.27*), expand the feature tree manager, and choose **Front Plane**.
6. Under **Translations**, click on the box for **Normal to Plane** (labeled 4 in *Figure 2.27*).

7. Click **OK**.

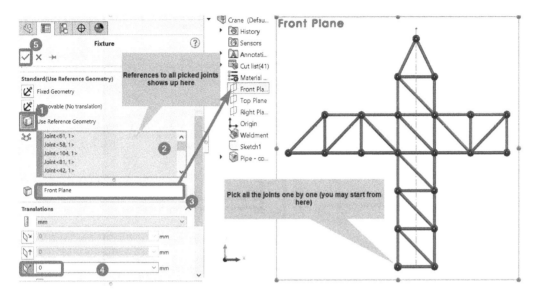

Figure 2.27 – Options for restraining the movements of all joints normal to the front plane

Steps 1-7 will impose a zero translational movement on all the nodes along the *z* axis. This ensures that we are doing a plane analysis.

Next, we will apply the restraints at joints **A** and **B**, both of which are located at the base of the crane, by following the steps given next.

Applying fixture on nodes A and B:

1. To apply the restraint on joint A, right-click on **Fixtures** and pick **Fixed Geometry**. Then do the following:

    I.   Under the **Fixture** property manager that appears, pick **Use Reference Geometry**.

    II. Move the cursor into the graphic window and pick only *joint A*.

    III. Choose **Front Plane** for the reference plane.

    IV. Under **Translation**, click on the arrows of the two boxes labeled 4 and 5 in *Figure 2.28 (a)*.

    V.   Click **OK**.

2. To apply the restraint on joint B, follow the options indicated in *Figure 2.28b*.

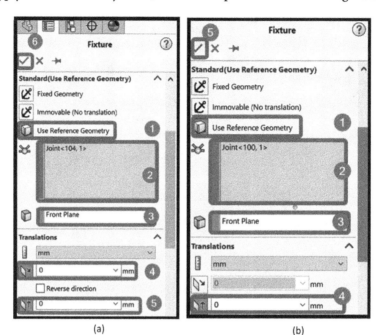

(a)   (b)

Figure 2.28 – (a) Options for restraining vertical and horizontal movements of joint A; (b) options for restraining vertical movement of joint B

Note that for joint *A*, we could also use the fixture named **Immovable (No translation)**; it performs the same function as what we did using **Use Reference Geometry**. For joint *B*, we restrained the movement along the vertical direction only, which is meant to replicate the behavior of a horizontal roller support.

At this stage, we are done with the application of all fixtures that need to be applied for our analysis. In the next sub-section, we will swing our attention to the specification of loads. This will inch us closer to running the analysis.

Applying external loads

Different types of loads can be used in SOLIDWORKS Simulation. For our analysis, we need to apply what is often referred to as payload weights represented by two vertical forces at joints R and W. We will apply the loads by using the **External Loads** command under the simulation study tree.

Follow these steps to apply the two forces at joints R and W, create the mesh, and then run the analysis:

1. Right-click **External Loads**.
2. Select **Force** (*Figure 2.29*).

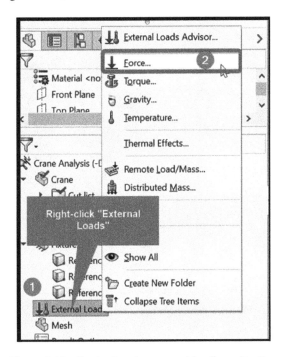

Figure 2.29 – Beginning the external load's application

3. From the **Force** property manager that appears, under **Selection**, click on the **Joints** symbol (labeled 1 in *Figure 2.30*).
4. Navigate to the graphic window and pick **joint R** (the leftmost joint).
5. Click inside the box labeled 3, then expand the feature manager tree to select **Front Plane** as the reference plane.

6. Under **Units,** keep it as **SI.**

7. Under **Force,** click in the second force component box (vertical force) and type 1500000.

8. Check **Reverse direction** (labeled 6 in *Figure 2.30*). This changes the direction of the force to a downward direction.

9. Click **OK.**

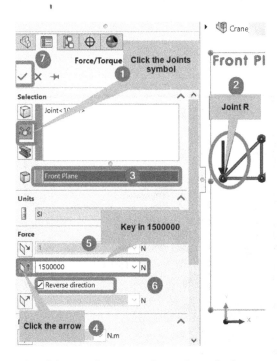

Figure 2.30 – Selecting the options for applying the force at joint R

10. Repeat *steps 1-9* to apply the 2000 kN force on joint *W* (in *step 4*, select joint *W*, and in *step 9*, key in 2000000).

After completing *step 10*, the appearance of the model in the graphics window will be as shown in *Figure 2.31*. The model is displayed in an isometric mode so that the arrows indicating the loads at joints *R* and *W* and the fixtures at all joints become apparent.

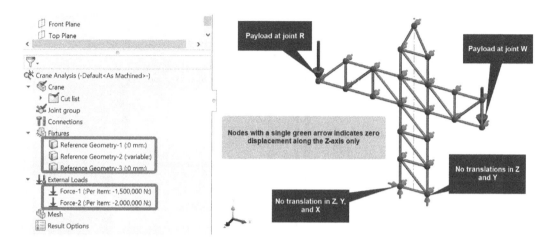

Figure 2.31 – The appearance of the model in the graphics window after applying loads and fixtures

The next task is to mesh and then run the analysis, which is what happens in the next sub-section.

Meshing

Meshing is an essential part of finite element simulation. There are two approaches for creating a mesh in SOLIDWORKS Simulation. The first involves creating the mesh using the command **Create Mesh**, while the second involves using the command **Mesh and Run**. For this chapter (as well as *Chapter 3, Analyses of Beams and Frames,* and *Chapter 4, Analyses of Torsionally Loaded Components*), we will be using the second approach. Principally, this approach combines the meshing and running of the analysis in a single step, and it works well whenever we use the weldment tool to create the members of a structure to be analyzed. Now, it is good to be aware that SOLIDWORKS does not provide the option for controlling the mesh quality for a structure idealized as a collection of **truss elements**. This means there is no point in engaging ourselves in the refinement of the mesh that we create for this problem. Further, it means if a truss is made of up 41 members, then only 41 **truss elements** are sufficient to analyze it accurately. Nonetheless, we will explore meshing in more detail, for instance, in *Chapter 5, Analyses of Axisymmetric Bodies,* and *Chapter 6, Analyses of Components with Solid Elements*.

Bearing the aforementioned detail in mind, we can now deal with the last steps before getting our desired results, to this end:

1. Right-click on **Mesh**.

2. Select **Mesh and Run** (as indicated in *Figure 2.32 (a)*).

After completing *steps 1 - 2*, the study tree will appear as shown in *Figure 2.32 (b)*.

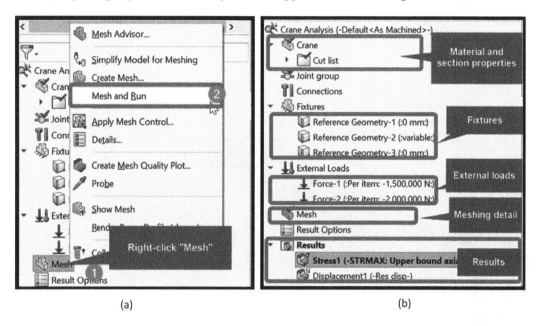

(a)   (b)

Figure 2.32 – (a) Creating the mesh and running the analysis; (b) Changes in the study tree after the Mesh and Run command

We will examine the results folder further in the next section.

Part D – Scrutinizing the results

Now that we have completed the steps in the previous sections (that is, *Parts A-C*), we are now in a position to make sense of the results and answer the following questions:

*   What is the maximum resultant deformation of the truss upon the application of the loads?

*   What is the distribution of the factor of safety of the members of the crane upon loading?

*   What is the internal force/stress that developed in the member IH?

But before answering these questions, it is worth noting that by default, when you use SOLIDWORKS for static studies, it generally computes, among others, the **Displacements** at the joints or nodes of the structure, the **Reaction forces** at the points of supports, the **Strains/Stresses** on an element/at the nodes, the **Factor of safety**, and so on.

Nevertheless, SOLIDWORKS Simulation will not always display all results. In fact, in most cases, the default results may not be what you want (you may simply right-click on them and then delete them). However, you can create custom plots of many more results and have them displayed in the Results folder. Let's start by examining the maximum resultant deformation in the next sub-section.

Obtaining the maximum resultant deformation

To retrieve the maximum resultant deformation experienced by the structure, simply navigate to the Results folder and double-click on **Displacment1 (-Res disp-)**, which is the result relating to the resultant displacement. After double-clicking, the graphics window with the results of the resultant displacement is shown in *Figure 2.33*.

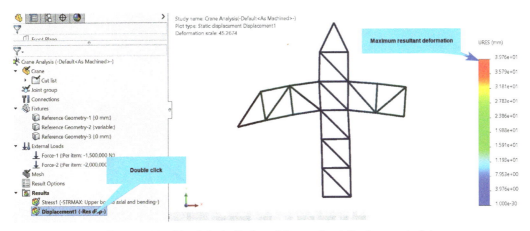

Figure 2.33 – The default display of the resultant displacement plot

The legend describing the displacement plot (on the far right of the screen) displays the range of the displacement from a very low number (in blue) to the maximum value (in red). It also displays the legend with the **scientific** number format. It is always better to present the result in a more readable format. Therefore, in the next set of steps, we will edit the display of the maximum resultant displacement value by following the steps given next:

1. Right-click on **Displacement1 (-Res Disp-)** and then choose **Edit Definition**.
2. From the **Displacement plot** property manager that appears, navigate to the **Chart Options** tab.
3. Under **Display Options**, tick inside **Show max annotation.**
4. Under **Position/Format,** change from **scientific** to **floating**.
5. Click **OK**.

Figure 2.34 indicates that with a combined load of 3500 kN applied to the crane, the maximum deformation experienced by one of the joints is **39.764 mm**.

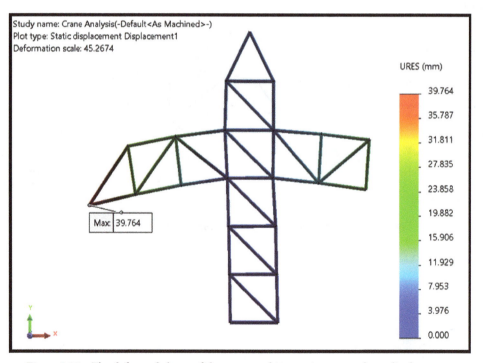

Figure 2.34 – The deformed shape of the crane and its maximum resultant displacement

Note that since a **truss element** has three translational displacement degrees of freedom at its node, the resultant displacement refers to the vectorial resultant displacement.

Obtaining the factor of safety

The **factor of safety** (**FOS**) is one of those results that are not automatically displayed but need to be retrieved during the post-processing of the results. From knowledge of the mechanics of materials, we know that the calculation of the factor of safety is based on certain failure criteria. SOLIDWORKS Simulation offers four failure criteria that we shall explore further in *Chapter 5*, *Analyses of Axisymmetric Bodies*. Nonetheless, irrespective of the criterion used in calculating the FOS, the rule of thumb is that the component we are analyzing has failed if the FOS is less than 1. But if the FOS is greater than 1, then the component is considered safe to support the applied load without failing (all things being equal). To retrieve the FOS, in the simulation study tree, do the following:

1. Right-click on the Results folder.
2. Select **Define Factor Of Safety Plot** (*Figure 2.35 (a)*).
3. The **Factor of Safety** property manager appears. Leave all options as they are as indicated in *Figure 2.35 (b)*. Notice that the failure criterion is set to **Automatic**. Also take note of the navigating arrow that unveils further options.
4. Click **OK**.

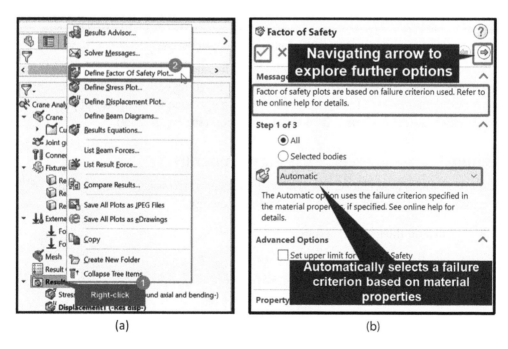

(a)   (b)

Figure 2.35 – (a) Retrieving the factor of safety; (b) the distribution of the factor of safety in the graphics window

Figure 2.36 reveals the distribution of the FOS for the crane upon the application of the load.

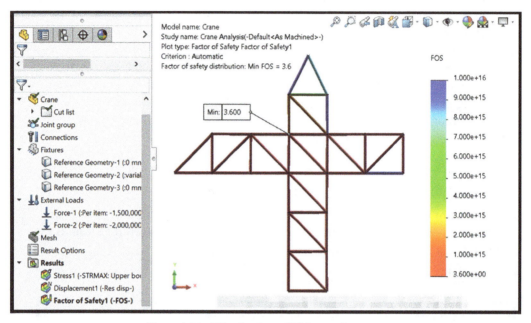

Figure 2.36 – Distribution of FOS over the crane

The figure indicates that the minimum FOS within the members of the crane is around 3.6, which means all is well with the crane. Note that by going with the **Automatic** option in step 3, SOLIDWORKS will use the failure criterion specified in the material property database, which is the **Max Von-Mises** failure criterion (see *Figure 2.23*).

Obtaining the axial force/stress for member IH

The last set of results we will look at is the axial forces and stresses. To retrieve either of these results, in the simulation study tree, do the following:

1. Right-click on the Results folder.

2. Select **List Beam Forces** (*Figure 2.37 (a)*).

3. Leave the options as indicated in *Figure 2.37 (b)*.

4. Click **OK**.

> **Note**
>
> To retrieve the axial stresses, change to **Stresses** in the box labeled 1 in *Figure 2.37 (b)*.

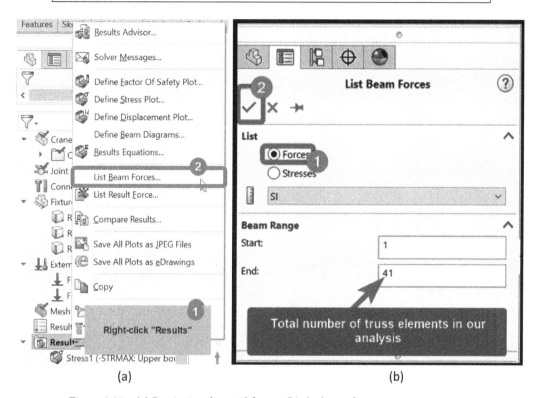

(a)   (b)

Figure 2.37 – (a) Retrieving the axial forces; (b) the beam forces property manager

Immediately after we complete steps 1-4, the **List Forces** window will appear in the graphics window as shown in *Figure 2.38*. You may use the arrow (marked 2) in *Figure 2.38* to navigate through the values of the axial force developed in different members of the structure. You can also click on the name of a specific member and SOLIDWORKS will instantly highlight it in the graphics window.

For instance, **Beam-19** (be aware that the name of this member will likely be different in your case) is the element that corresponds to member **IH**. It has been highlighted within the **List Forces** window for convenience. As you can see, this element experiences an internal axial force of approximately **727 kN**. This value is within 3% of the answer (**707 kN**) obtained in *[1]*. Note that this is the only value computed in *[1]*, simply because it is not trivial to carry out the manual calculations of the other values (displacements, FOS, forces, and stresses) without incurring substantial errors. Indeed, the small difference between the value from SOLIDWORKS and the cited reference may be attributed to possible rounding off errors.

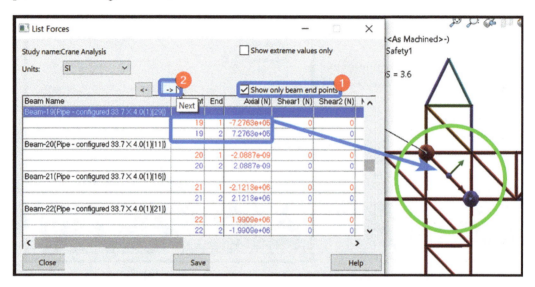

Figure 2.38 – The default display of axial forces

In many practical instances, you will want to list out all the forces/stresses for further examination. To do this, simply click on the column name **Axial (N)** (labeled 1) in *Figure 2.39*. This will give a compact display of the values of all 41 elements used in our analysis. From here, click on the **Save** button at the bottom of the **List Forces** window (labeled 3) to save all the data to an Excel file, then click **Close** to close the window. Afterwards, you can explore the data in Microsoft Excel.

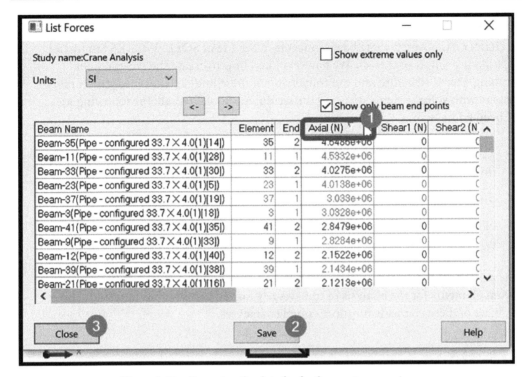

Beam Name	Element	End	Axial (N)	Shear1 (N)	Shear2 (N
Beam-35(Pipe - configured 33.7 X 4.0(1)[14])	35	2	4.6486e+06	0	(
Beam-11(Pipe - configured 33.7 X 4.0(1)[28])	11	1	4.5332e+06	0	(
Beam-33(Pipe - configured 33.7 X 4.0(1)[30])	33	2	4.0275e+06	0	(
Beam-23(Pipe - configured 33.7 X 4.0(1)[5])	23	1	4.0138e+06	0	(
Beam-37(Pipe - configured 33.7 X 4.0(1)[19])	37	1	3.033e+06	0	(
Beam-3(Pipe - configured 33.7 X 4.0(1)[18])	3	1	3.0328e+06	0	(
Beam-41(Pipe - configured 33.7 X 4.0(1)[35])	41	2	2.8479e+06	0	(
Beam-9(Pipe - configured 33.7 X 4.0(1)[33])	9	1	2.8284e+06	0	(
Beam-12(Pipe - configured 33.7 X 4.0(1)[40])	12	2	2.1522e+06	0	(
Beam-39(Pipe - configured 33.7 X 4.0(1)[38])	39	1	2.1434e+06	0	(
Beam-21(Pipe - configured 33.7 X 4.0(1)[16])	21	2	2.1213e+06	0	(

Figure 2.39 – Preparing the data for further post-processing

You will notice that beyond the axial forces, the **List Forces** window has other columns, such as **Shear1**, **Shear2**, and so on. At the moment, they all are zeros because we are employing the **truss element** option. For analyses that require the use of beam elements such as frames, they will not be zero.

Other things to know about the truss element

We have used the truss element in this chapter to study the computer analysis of a crane idealized as a 2D planar truss. However, the truss element and the procedure outlined in this chapter are also perfectly suitable to study straight bars and 3D space trusses. For conciseness, the deployment of the truss element to investigate the structural performance of components made of simple bars is demonstrated in the computer file of the first exercise question at the end of this chapter.

Summary

This chapter has covered some basic concepts in the use of SOLIDWORKS Simulation for the static analysis of trusses. We have explored how to create the skeletal lines describing a truss structure and the conversion of these line sketches into structural members with volume elements using the weldments tool. Overall, the following ideas have been addressed:

- How to activate weldment tool for structural analysis
- How to edit the cross-section of in-built profiles for specific needs
- How to change structural beam elements to truss elements
- Applying loads and fixtures on specific joints of a truss structure
- Modifying the displacement plot and obtaining the factor of safety

In the next chapter, *Chapter 3, Analyses of Beams and Frames*, we will study the use of **beam elements** for the analysis of transversely loaded components and study the usefulness of these elements for more complex analyses.

Questions

1. *Figure 2.40* shows two straight segments of a machine loaded as shown. The segments are made of alloy steel and have a cross-sectional profile with an external and internal diameter of *40 mm* and *20 mm*, respectively. Treat the segments as two connected bars, then use SOLIDWORKS Simulation to do the following:

    a. Determine the displacement of end C.

    b. Evaluate the axial stresses developed in the components upon loading.

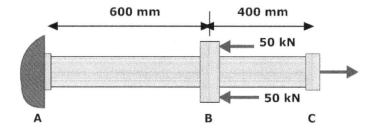

Figure 2.40

2. *Figure 2.41* shows a 2D plane truss representing a load-supporting mechanism. Components CB and AB are made of ASTM A-36 steel tubes with the same cross-section (external and internal diameters of *50 mm* and *30 mm* , respectively). Use SOLIDWORKS Simulation to do the following:

a. Determine the resultant displacement of joint B

b. Determine the minimum factor of safety of the assembly

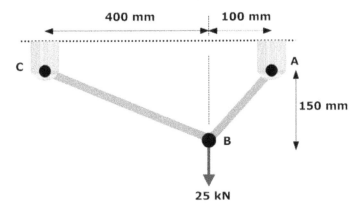

Figure 2.41

Further reading

[1] *Structural and stress analysis, T. H. G. Megson, 4th, Ed, Butterworth-Heinemann, 2019.*

3
Analyses of Beams and Frames

This chapter focuses on the SOLIDWORKS Simulation procedure and strategy for the analysis of transversely loaded members, mainly what we call beams and frames. In our treatment of these structures, we will uncover a few more details about SOLIDWORKS weldment tool, examine how to apply more complex types of load (concentrated load, moment, and distributed load), and explore further computed results that relate to beams. At the end of this chapter, you will become familiar with the procedure and tricks for the simulation of the aforementioned structures. In this vein, the chapter is organized around the following topics:

- An overview of beams and frames
- Strategies for the analysis of beams and frames
- Getting started with analyzing beams and frames in SOLIDWORKS Simulation

Technical requirements

You will need to have access to SOLIDWORKS' software with a SOLIDWORKS Simulation license.

You can find the sample files of the models required for this chapter here: `https://github.com/PacktPublishing/Practical-Finite-Element-Simulations-with-SOLIDWORKS-2022/tree/main/Chapter03`

An overview of beams and frames

This section provides basic background information about beams and frames. It highlights their applications, the objectives of simulating/analyzing these structures, and some important technical features relevant to their analysis.

You will likely be familiar with the technical jargons of the *mechanics of Solids* and its restrictive definition of a **simple beam** as a structure that satisfies these conditions:

- Capable of supporting transverse loads, which are loads that are applied perpendicular to its length

- Has a length that is far greater than the dimensions of its cross-sections

However, in a more general sense, a beam can support transverse, axial, and torsional loads. But in this chapter, we will only focus on beams supporting transverse loads.

Beams remain one of the most flexible categories of structures used for the design of many engineering products, machines, and systems. You will find them in simple forms, such as a diving board, parallel bars for fitness training, balance beams in gymnastics, lintels that support windows and doors in buildings, and so on. They also exist in more complex forms as part of heavy machinery, medical equipment, vehicles, lifting platforms, buildings, bridges, and so on.

Lengthwise, beams may be designed as a single span, that is, as one long member (as shown in *Figure 3.1a*), or as a set of continuously connected members (*Figure 3.1b*). Just like the truss structure we treated in *Chapter 2, Analyses of Bars and Trusses* (which is a collection of bars arranged to form a larger assembly), we can also bring together a collection of beams and then arrange them to form a bigger assembly that becomes what is known as **frames** – as reflected in *Figure 3.1c* and *Figure 3.1d*:

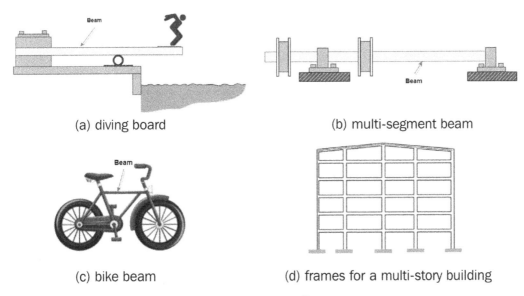

(a) diving board   (b) multi-segment beam

(c) bike beam   (d) frames for a multi-story building

Figure 3.1 – Some applications of beams in practice

When we conduct static analyses of beams and frames, our objectives may center around the following:

- To determine the internal forces and their variations along the length of the members.

- To determine the stresses that developed within the structure or its members (if made up of more than one distinct segment). The stress may be either normal stress caused by the bending moment or shear stress induced by shear forces.

- To evaluate the deformation of the beam or its members (if made up of more than one beam as in the case of connected beams or a frame). A simple beam deformation is characterized by deflection and slope.

- To determine the reaction forces.

As is always the case, the purpose of each of the preceding objectives is to ensure that the beam can support the loads applied to it without premature failure or excessive deformation that will lead to its instability/inability to carry the load.

> **Important Note**
>
> More details about internal loads, types of support, and other details about beams can be found in books on mechanics of solids, such as *Hibbeler [1]* (see the *Further reading* section), *Bedford* and *Liechti [2]*, and others. Nevertheless, recall that if a beam is loaded only by transverse loads, then bending moments and shear forces are the major internal resistant forces in these structures. In a similar spirit, if a beam is loaded by a combination of transverse, axial, and torsional loads, then the internal forces will include bending moments, shear forces, axial, and torsional resistance loads.

The following technical points are worth noting about beams:

- Beams may be long and straight in form, but they may also be curved. However, we will only deal with straight beams in this chapter.

- A beam's cross-section may be uniform or non-uniform across its length. Simple cross-sections of beams are solid/hollow rectangular sections or solid/hollow circular sections. But the cross-section can also be any of the other standard types shown in *Figure 3.2a*.

- Some beams will have symmetric cross-sections, while others may have non-symmetric cross-sections. We will focus only on symmetric cross-sections in this chapter.

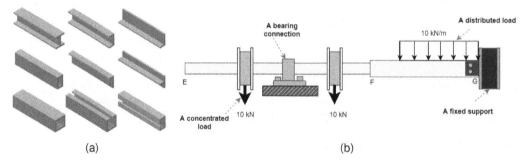

(a)   (b)

Figure 3.2 – (a) Some typical beam cross-sections; (b) An example of two connected beams with loads and supports

- Some common examples of transverse loads include concentrated forces/moments (acting at a point) and distributed forces/moments (occupying a segment). *Figure 3.2b* depicts two concentrated forces and a distributed force.

With some of this background information provided about beams, we will briefly highlight strategies for their simulations in the next section.

Strategies for the analysis of beams and frames

This section describes the structural details that are necessary for the analysis of beams and frames. It also highlights the modeling strategy for their simulation and the characteristics of the **beam element** within the SOLIDWORKS Simulation library.

Structural details

You will need to know the following technical details before venturing into the analysis of beam/frame structures:

- Dimensions

    a. The details of the cross-section

    b. The geometric length of each member

    c. The orientation angles of the members (for a frame)

- The material properties of the beam/members of a frame

- The external loads applied to a beam/frame

- The support provided to a beam/frame to prevent a rigid body motion

Beyond the structural details, we are now set to examine the modeling strategy.

Modeling strategy

Armed with the geometric and material details, the steps to take for the analysis of beams/ frames are depicted in *Figure 3.3*:

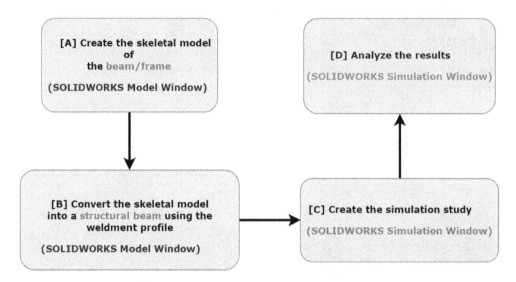

Figure 3.3 – Main steps for the static analysis of beams and frames

There is a subtle difference between this figure and that in *Chapter 2, Analyses of Bars and Trusses*. Mainly, in step A, we are creating the skeletal model of a beam/frame structure to be simulated. Now, because we will be using the **weldments** tool, there is an important point you need to know. Primarily, in creating the skeletal line model of a beam-/frame-based structure, you should always consider that there is a virtual joint at critical positions along the length of the beam/frame. These critical positions are located at the following points:

- The beginning and end of a beam
- A point where a concentrated load is applied
- The beginning of a distributed load
- The end of a distributed load
- A point where there is a change of cross-section or change of material properties
- All points of support
- All points of connections

In many practical scenarios, some of these critical points may not be what you would regard as a joint. For instance, to model the beam in *Figure 3.4a*, a good strategy is to assume the critical positions as shown in *Figure 3.4b* and then create each of the line segments (*AB, BC, CD, DE, EF,* and *FG*) separately before invoking the weldment tool. In a similar spirit, to model the frame structure in *Figure 3.4c*, we will assume the critical positions as shown in *Figure 3.4d* and create the necessary line segments (*AB, BE, EF, FC,* and *CD*) accordingly:

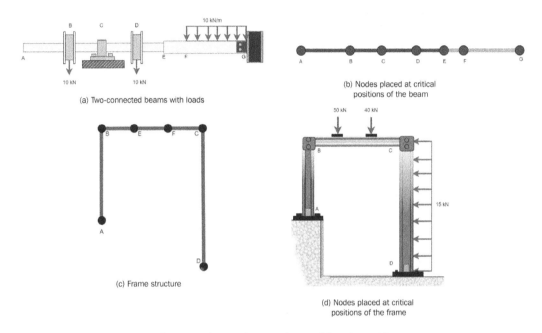

(a) Two-connected beams with loads

(b) Nodes placed at critical positions of the beam

(c) Frame structure

(d) Nodes placed at critical positions of the frame

Figure 3.4 – Indication of critical points for modeling beam/frame structures

Notice that *points of supports* and *connections* are mentioned among the critical positions listed previously. This indicates that during modeling, the strategy for dealing with different types of support or connections is not to model them directly. Instead, they are represented as joints.

Furthermore, be aware that various types of simple or advanced connections (such as welding, riveting, ball and socket joints, simple thrust bearings, and so on) may be involved in components to be analyzed. A common modeling strategy is to approximate these joints as either *fixed/clamped*, *hinge/simply supported*, or *roller/sliding support* during analysis. Nonetheless, a sound technical judgment of which type of approximation to assign to a connection is necessary.

You will learn more about SOLIDWORKS Simulation features that are used as proxies for each of these types of support in a later segment of this chapter – *Part C: Create the simulation study*.

With coverage of the modeling strategies out of the way, it is time to reiterate the characteristics of the beam element that we will employ during the simulation case study of this chapter.

Characteristics of the beam element in the SOLIDWORKS Simulation library

Within the SOLIDWORKS Simulation environment, the beam element is used to analyze both beam and frame structures. Like the truss element in *Chapter 2, Analyses of Bars and Trusses*, the beam element has two nodes. However, there is one major difference. The beam element has six degrees of freedom per node because it is a three-dimensional beam structure. The six degrees of freedom encompass the following:

- Three translational displacements about the *x*, *y*, and *z* axes
- Three rotational displacements about the *x*, *y*, and *z* axes

So far, we have learned about the modeling strategies and the features of the beam element. It is therefore time to go through a case study to explore further details about simulating with this element.

Getting started with analyzing beams and frames in SOLIDWORKS Simulation

In this section, we will demonstrate the use of the beam element with a practical case study that exemplifies a beam carrying multiple types of load.

Through this case study, you will become familiar with how to apply transverse concentrated forces and concentrated moments at a joint. You will also learn how to apply a distributed load on a beam segment. Finally, you will get to see how to extract shear force and bending moment diagrams (which are simulation results that are specific to beams/frames analysis).

Time for action – Conducting a static analysis of a beam with multiple loads

Problem statement

We will analyze the structural component that is loaded and supported as shown in *Figure 3.5a*. The beam is made of AISI 304 steel and has an S (American Standard) cross-section – S 120 x 12, which is the cross-sectional profile depicted in *Figure 3.5b*. Using SOLIDWORKS Simulation, we want to accomplish the following tasks:

• Determine the maximum vertical deflection of the beam and the location of this maximum deflection.

• Visualize the variation of shear force and bending moment along the length of the beam.

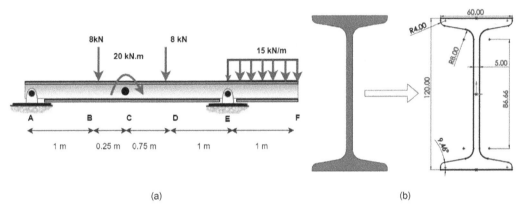

(a)   (b)

Figure 3.5 – (a) A 2D schematic of a single span beam with multiple loads; (b) An S cross-section profile – S 120 x 12 (unit in mm)

This case study is inspired by a practical problem suggested by *Hibbeler [3]*. By the end of this simulation study, we will compare the simulation solution we obtained with the partial results provided in *[3]*.

But first, let's commence the simulation study by initiating the creation of the beam's model.

Part A – Create a sketch of lines describing the centroidal axis of the beam

As is always the case, creating the geometric model of the structure to be analyzed is the first step in SOLIDWORKS Simulation. in this vein, this section demonstrates the steps to create the geometric line denoting the centroidal axis of the beam.

To begin, start up SOLIDWORKS (**File → New → Part**) and then create five line segments, *AB*, *BC*, *CD*, *DE*, and *EF*, by following the steps highlighted next. Take note that we will sketch the line model of the beam on the front plane.

Let's get started:

1. Click on the **Sketch** tab.
2. Click the **Sketch** tool.
3. Choose **Front Plane,** as shown in *Figure 3.6*:

> **Important Note**
>
> In structural analysis, there is often a long discussion about the difference between the neutral axis and centroidal axis. The reference to the centroidal axis in this section is used to indicate the line passing through the center of mass of the beam we are analyzing.

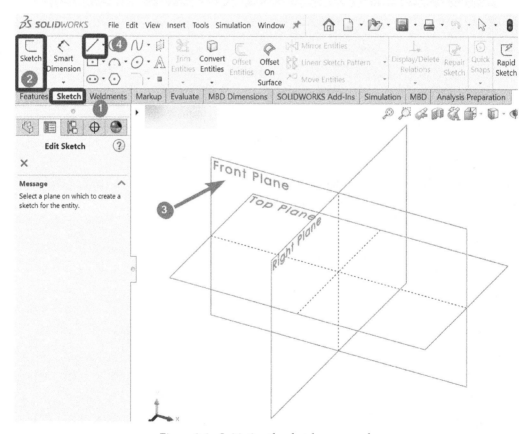

Figure 3.6 – Initiating the sketch command

4. Select the **Line** sketching command and create the lines based on the dimensions in *Figure 3.7*.

5. Click on the **Exit Sketch** symbol, as indicated in *Figure 3.7*. Note that exiting the sketch will not close it; it only deactivates the sketching mode.

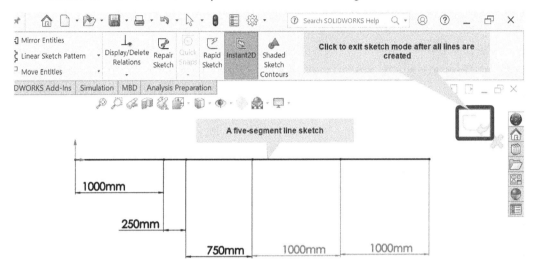

Figure 3.7 – The sketched line segments

The sketched line we have created represents the centroidal axis of the beam. However, the sketched line has no volume property yet (as we discussed in *Chapter 2, Analyses of Bars and Trusses*).

In the next section, our objective is to convert the sketched line into a structural model with a volume property.

Part B – Convert the skeletal model into a structural profile

Here, we will convert the sketched lines we created in the preceding section into a structural model using the weldments tool that was introduced in *Chapter 2, Analyses of Bars and Trusses*.

Activating the weldments tool

Check the set of command items in your SOLIDWORKS **CommandManager** tabs to see whether the **Weldments** tab is present. If it is present, then skip the steps here and move to the next sub-section (*Adding a structural property*).

But if the **Weldments** tab is absent from the set of **CommandManager** tabs, then follow the steps highlighted next (summarized in *Figure 3.8*):

1. Right-click on the **CommandManager** tab to bring up the option to show more tabs.

2. Move your cursor over the word **Tabs**.

3. Click on **Weldments**.

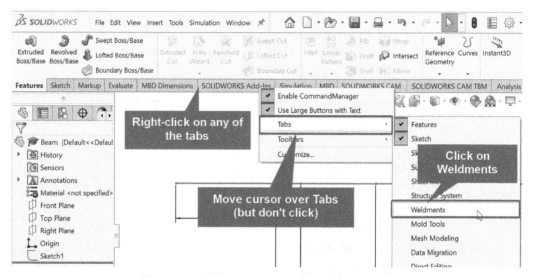

Figure 3.8 – Options to activate the Weldments tab

Once steps 1–3 are complete, the **Weldments** tab will appear, and we can then make use of the **Structural** command, as explained in the next subsection.

Adding a structural property

We need to add the cross-sectional profile to the line segments created in the preceding subsection to give it a volume property.

To do this, perform the following steps

1. Click on the **Weldments** tab, as shown in *Figure 3.9*.

2. Click on **Structural Member**.

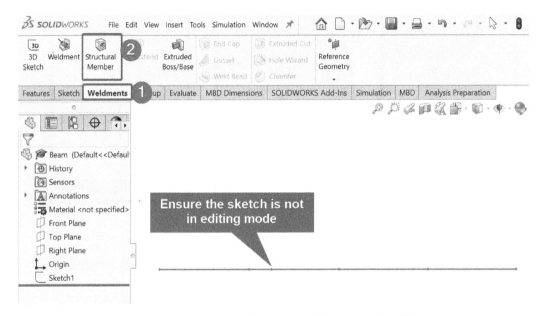

Figure 3.9 – Beginning the addition of a structural profile

After completing *step 2*, the **Structural Member PropertyManager** window appears, and we now have to make selections of the desired profile from the Simulation library, as documented in *Figure 3.10*.

3. Under **Standard**, choose **iso** (*Figure 3.10a*).

4. Under **Type**, select **sb beam** (*Figure 3.10b*).

5.   Under **Size**, select **120 x 12** (*Figure 3.10c*).

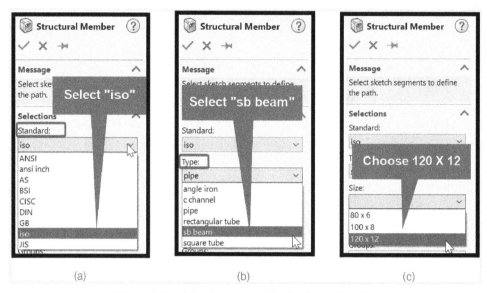

Figure 3.10 – Selection options for the desired profile

We still have a few more steps before completing the task in this subsection. Hence, do not close **Structural Member PropertyManager** yet.

6.   Within **Structural Member**, click **New Group** to form **Group 1** and select all five lines, as shown in *Figure 3.11*:

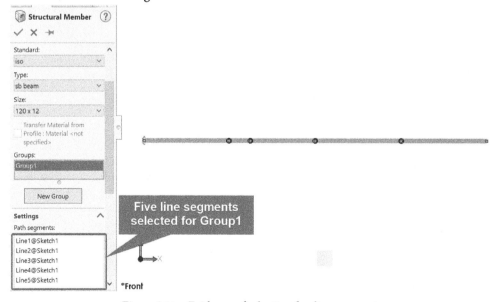

Figure 3.11 – Evidence of selecting five line segments

Before we finish the selection, it is a good idea to check the orientation of the profile to be sure it is what we want.

7. To check and reorient the profile, move to the graphics area and zoom in on the cross-section profile. It is likely to appear as shown in *Figure 3.12a*. If this is the case, then we have to reorient the profile to stand vertically.

8. Click inside the **Rotation Angle** box and input 90 deg. This will rotate the profile as demonstrated in *Figure 3.12b*.

9. Click **OK**.

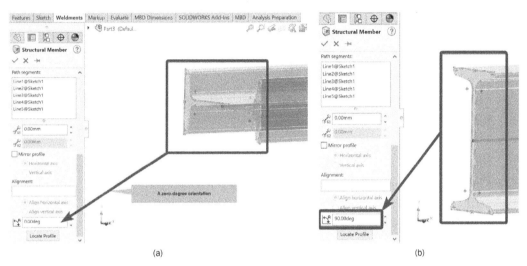

(a)   (b)

Figure 3.12 – (a) Default orientation of the profile; (b) Rotation of
the cross-section to the desired orientation

Important Note

Note that it is always necessary to check the orientation of the profile for beam structures as we have done previously. The main reason for this is hinged on the fact that the performance analysis of beam-based structures depends on the second moment of inertia – a geometric parameter that depends on the orientation of the cross-section.

After completing *steps 3–9*, *Figure 3.13* shows a partial view of the solid beam and the changes to the **FeatureManager** tree:

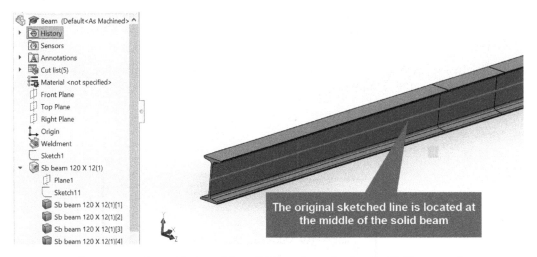

The original sketched line is located at the middle of the solid beam

Figure 3.13 – A partial view of the solid beam formed using the Weldment tool

Now that we have converted the sketched line into a collection of solid bodies with a volume property, it is time to bring forth the tools for the analysis.

Part C – Create the simulation study

This section comprises several subsections, each dealing with distinct features of the simulation tasks. The first sub-section deals with the activation steps for the simulation study. This is then followed by a specification of the material for the beam. After this, we will drill down into the application of fixtures and loads. The final subsection relates to the meshing of the structure along with the running of the study to obtain the results.

Let's go ahead and create the study:

1. Activate the **Simulation** tab and create a new study.

2. Click on **SOLIDWORKS Add-Ins**.

3. Click on **SOLIDWORKS Simulation** to activate the **Simulation** tab (see *Figure 3.14*).

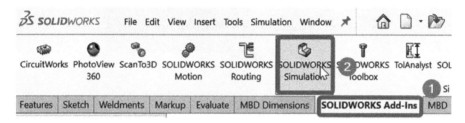

Figure 3.14 – Activating the Simulation tab

4. With the **Simulation** tab active, click on **New Study** (see *Figure 3.15*).

Figure 3.15 – Activating a new study

Study PropertyManager appears after *step 4* is completed. Within **Study PropertyManager**, perform the following steps.

5. Input a study name within the **Name** box, such as `Beam analysis`.

6. Keep the **Static** analysis option (as shown in *Figure 3.16*).

7. Click **OK** (that is, the green checkmark).

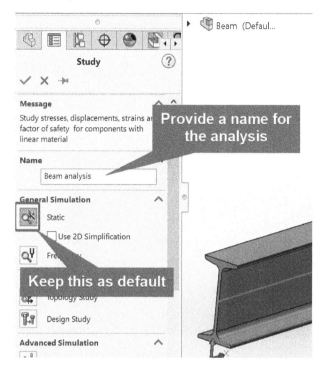

Figure 3.16 – Specifying options for Study PropertyManager

Upon completing *steps 4–6*, two changes will happen:

- The simulation study tree becomes active (*Figure 3.17*).

- Joints will be created at the intersection of the beam members and the free ends.

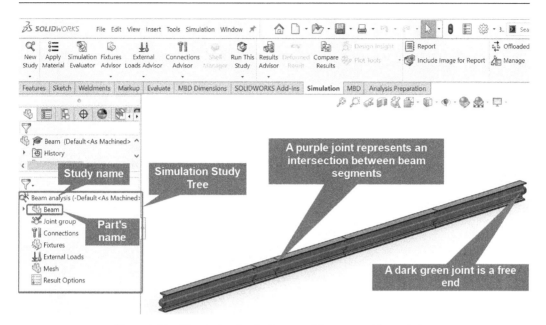

Figure 3.17 – Simulation Study PropertyManager and beam joints

Take note of the two colors used for the joints of the beam depicted in the graphic window of *Figure 3.17*. The joints in purple represent connections between beam segments, while the dark green joints imply the free ends of the beam.

In the next subsection, we will supply details of the material properties of the beam.

Adding a material property

According to the problem statement, the beam is made of **AISI 304 steel**, so let's define the beam's material property by following these steps:

1. Under the Simulation study tree, right-click on **Beam**. This brings up the option to select and apply material properties, as shown in *Figure 3.18*:

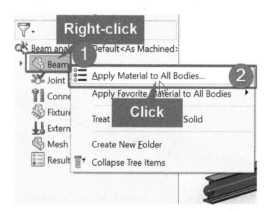

Figure 3.18 – Activating the material database

2. Click **Apply Material to All Bodies**.

    Notice that *step 2* launches the material database depicted in *Figure 3.19*. With the **Materials** window open, perform the following steps.

3. Look for the **Steel** folder.

4. Click on **AISI 304** (see *Figure 3.19*).

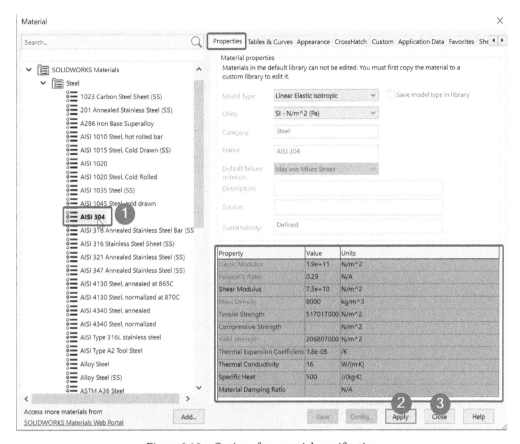

Figure 3.19 – Options for material specification

Once you click on **AISI 304**, its properties will be revealed on the right side of the **Material** window. From this window, you will observe that some of the property names are in blue font while others are in red or black font (see *Figure 3.19*). In general, the material properties with names in red font are the ones that must be specified for static analysis. The other property names in blue or black font are either optional for static analysis or needed for dynamic and thermal analyses.

With this brief detail regarding the **Material** window, let's now wrap up the material specification steps:

5. Click **Apply**.
6. Click **Close**.

After closing the material database window, a green tick mark (✓) will appear on the part's name (which we have already explored in *Chapter 2, Analyses of Bars and Trusses*).

Having completed the material specification steps, it is time to move on to the specification of the beam element to be used for our simulation.

Confirming the beam element status for the segments

SOLIDWORKS Simulation treats a structural member that is created using the weldment tool as a beam by default. Consequently, the software employs the beam element for the simulation of such a structure. The purpose of this subsection is to simply confirm this.

For the confirmation, in the Simulation study tree, do the following:

1. Click on the arrow beside the part's name to expose the folders named **Cut list** (see *Figure 3.20*).

2. Expand the **Cut list** folder to expose the subfolders named **SB BEAM 120.00 x 12.00**.

3. Select all the beam segments (each named **SolidBody…**).

4. Right-click on one of the selected segments, and then choose **Edit Definition**.

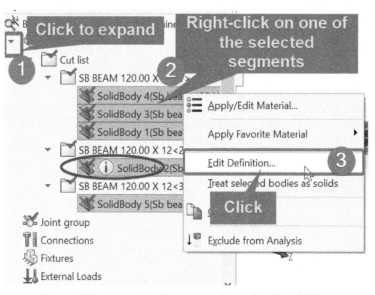

Figure 3.20 – Revealing the structural members' subfolders

Looking closely in *Figure 3.20*, you will notice that a beam segment (enclosed by a purple ellipse) has a warning sign (ⓘ) attached to its name. This warning is about the shortness of this beam segment and it can be ignored without any issue. We will explore the warning further in *Chapter 4, Analyses of Torsionally Loaded Components.*

After completing *steps 1–4*, **Apply-Edit Beam PropertyManager** will appear:

1. Keep the selection as shown in *Figure 3.21*.

2. Click **OK** (the green tick mark) to accept all the default options.

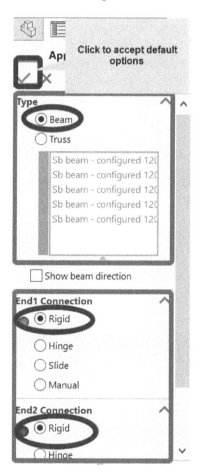

Figure 3.21 – Confirming the beam element selection

You will notice that **Beam** is selected by default, as shown in *Figure 3.21*. Furthermore, you will also observe that within **Apply-Edit Beam PropertyManager**, there are a few possible options to specify the nature of **End1 Connection** and **End2 Connection**.

However, the default option that we have gone with here is the **Rigid** connection for both the **End1** and **End2** connections. The rigid option indicates that no forces or moments are released at the ends of each beam segment. It is generally the safest option. Proxies for the other options can be specified using the fixture command. Coincidentally, the next section will take us through the process of applying a fixture.

Applying a fixture

Fixtures are used to prevent engineering structures from excessive unstable movement when loads are applied to them.

For the problem we are analyzing, the supports found at locations A and E (as indicated in the problem statement – *Figure 3.5a*) represent the fixtures. You will notice that the two supports are pin connections. Each pin connection will act to prevent the three translational movements along the x, y, and z axes at each joint.

We will apply the two fixtures in the same set of steps because they are of the same nature. Consequently, to apply the fixture at joints A and E, follow these steps.

Under **Simulation study tree**, do the following:

1. Right-click on **Fixtures**, and then pick **Fixed Geometry** from the context menu that appears (*Figure 3.22*).

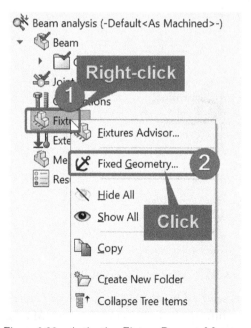

Figure 3.22 – Activating Fixture PropertyManager

2. Under **Fixture PropertyManager** (*Figure 3.23*), select **Immovable (No translation)**.

3. Move the cursor into the graphics window and pick joints *A* and *E*.

4. Click **OK**.

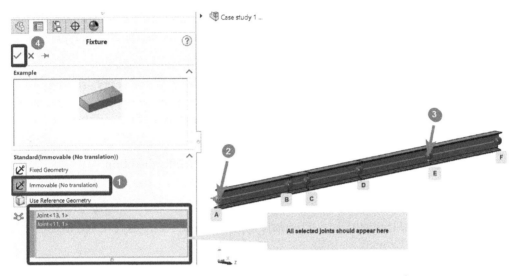

Figure 3.23 – Selecting the immovable fixture for joints A and E

Important Note

Note that in many other practical analysis cases, you may have a combination of support types that perform different functions, such as fixed support and a pin connection at different positions. In those cases, you will need to apply each fixture one at a time, not at once, as we have done previously by selecting joints *A* and *E* in one single set of steps.

It bears pointing out that SOLIDWORKS Simulation provides commands to apply the three most common types of supports when dealing with beams. Let's cycle through these prominent supports before moving on to the next subsection:

- *Fixed/clamped support* – This type of support prevents the three translational movements and the three rotational displacements when applied to a joint. In SOLIDWORKS Simulation, this type of support is handled by the fixture type called **Fixed Geometry**.

- *Pin/hinge/simply support* – This kind of support prevents all three translational movements at a joint. Within the SOLIDWORKS Simulation environment, this type of support is handled by the fixture indicated by **Immovable (No translation)**.

- *Roller support* – This type of support prevents a limited number of translational displacements of a joint. Within the SOLIDWORKS Simulation environment, we tackle this type of support by using the **Use Reference Geometry** fixture type.

Of course, apart from the three common support types mentioned previously, there are also others. This includes *elastic support* and *sliding support*. We can apply sliding supports on a beam's joint using the **Use Reference Geometry** fixture type. However, take note that SOLIDWORKS Simulation only allows the application of elastic support for a **solid/shell element**.

We now turn to the application of the external loads in the next subsection.

Applying loads

We have three types of loads that are externally applied to the beam, as revealed in the following problem statement:

- Concentrated payload weights at positions *B* and *D*

- A concentrated moment at position *C*

- A **uniformly distributed load** (UDL) occupying segment *EF*

The steps to apply to each of these loads are discussed next. We will start with the concentrated loads, then the moment load, and finally we will apply the UDL.

Applying the concentrated loads at joints B and D

Under **Simulation study tree**, perform the following steps:

1. Right-click on **External Loads**, and then select **Force** (as shown in *Figure 3.24*).

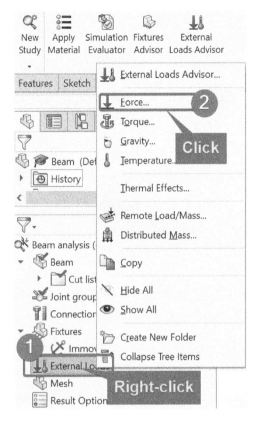

Figure 3.24: Activating Force/Torque PropertyManager

Within **Force/Torque PropertyManager**, perform the following steps.

2. Under **Selection**, click on the **Joints** symbol (labeled **1** in *Figure 3.25*).

3. Navigate to the graphics window and pick joint *B*.

4. Click inside the load reference box (labeled **3** in *Figure 3.25*).

5. Expand the **FeatureManager** tree to select **Front Plane** as the reference plane.

6. Under **Force**, activate the vertical force component box and input 8000 (ensure the unit is set to an **SI** unit).

7. Check **Reverse direction** (labeled 7 in *Figure 3.25*). This changes the force direction to downward.

8. Click **OK**.

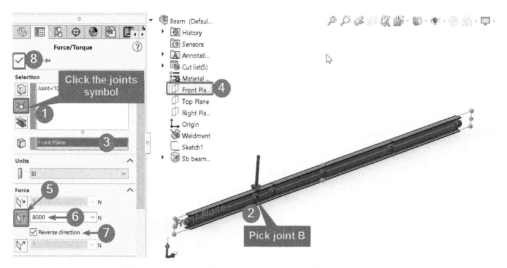

Figure 3.25 - Applying a concentrated force at joint B

You will notice that **Force/Torque PropertyManager** has five areas:

- *Selection* – This is used for specifying geometry on which a load is applied.

- *Units* – This is used for indicating the desired unit of the load we are applying.

- *Force* – This section is for specifying the magnitude and direction of a force.

- *Moment* – This section is for specifying the magnitude and direction of a moment.

- *Symbol Settings* – This is valuable for increasing the size of the arrow representing a load. Usually, the arrow that signifies a load may be too small. See *Figure 3.26* on how to modify the size and color of the arrow.

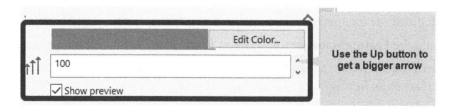

Figure 3.26 – Editing the size and color of a load's arrow symbol

Repeat *steps 1–8* to apply the 8000 N force on joint *D*. However, remember to select joint *D* in *step 3*.

Applying the concentrated moment at joint C

We are now set to apply the moment at joint *C*. Before we do that, there are three things to take note of when dealing with the load type called **moment**:

- First, in most engineering mechanics textbooks, it is typical to employ a curved arrow to represent an externally applied moment (as illustrated in *Figure 3.5a*). However, within SOLIDWORKS Simulation, the symbol used to represent a moment appears as a nail, as you will see later in this subsection.

- Second, SOLIDWORKS Simulation employs a convention that holds that a counterclockwise moment is positive, while a clockwise moment is negative.

- Third, a moment applied to a point will need a reference plane. Selecting the right reference plane for a moment is not as intuitive as that of a force. So, care should be taken in the steps involved in specifying the reference plane of the moment.

With the preceding background information, let's now shift our attention back to applying the moment at joint *C* by following the steps spelled out next.

Under **Simulation study tree**, do the following:

1. Right-click on **External Loads** and select **Force**.

2. Within **Force/Torque PropertyManager**, follow the options highlighted in *Figure 3.27*:

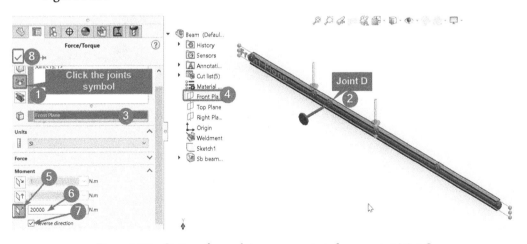

Figure 3.27 – Options for applying a concentrated moment at joint C

Notice that the action labeled 7 in *Figure 3.27* invokes the **Reverse direction** command. This step is necessary to ensure that the applied moment is negative in compliance with the problem statement. In addition to this, you will see that the symbol of the moment aligns with the *Z* axis.

As you have seen so far, concentrated forces and concentrated moments act at a joint. But we could also have loads that will act on a segment, rather than on joints. This load type is called a **distributed load** and it is the focus of our next action.

Applying the distributed load on segment EF

In the final step for this subsection, we will apply a distributed load on segment *EF* of the beam.

Under **Simulation study tree**, do the following:

1. Right-click **External Loads** and select **Force**.

2. Within **Force/Torque PropertyManager**, follow the steps highlighted in *Figure 3.28*:

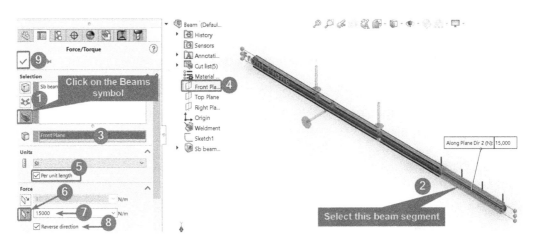

Figure 3.28 – Applying a distributed load on segment EF of the beam

After completing the preceding steps, the appearance of the model in your graphic window should be as shown in *Figure 3.29* (viewed isometrically). Notice the symbols representing the concentrated forces at points *B* and *D*. Also take note of the concentrated moment at point *C* and the distributed load on segment *EF* of the beam. Additionally, pay attention to the support at joints *A* and *E*.

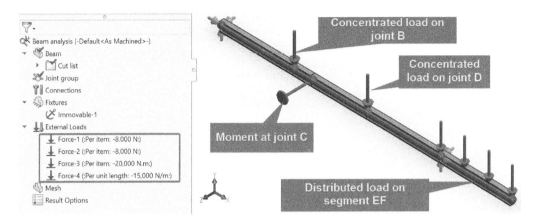

Figure 3.29 – Appearance of the different external loads

We have now completed the application of the three loads. We will be creating and running the mesh in the next subsection.

Meshing and running

We are getting close to completing the preprocessing steps. Meshing is a crucial step of the finite element simulation. As previously mentioned in *Chapter 2, Analyses of Bars and Trusses*, the two approaches for creating a mesh in SOLIDWORKS Simulation are **Create Mesh** and **Mesh and Run**. In this subsection, we will be using the second approach, which combines the meshing and running of the analysis in a single step. The method works well when analyzing components with either truss or beam elements. Be aware that when this method is used for the structure we are analyzing, it will create an element between two joints. Besides, we do not have to do detailed mesh refinement for this problem because of the strategic use of critical positions employed in creating the geometric model of the beam in the section named *Part A – Create a sketch of lines describing the centroidal axis of the beam*. As you will later see in *Chapter 5, Analyses of Axisymmetric Bodies*, and *Chapter 6, Analyses of Components with Solid Elements*, for more advanced analysis, a mesh convergence study will be crucial to get accurate results.

For our next action, perform the following steps:

1. Right-click on **Mesh**.
2. Select **Mesh and Run** (as highlighted in *Figure 3.30*).

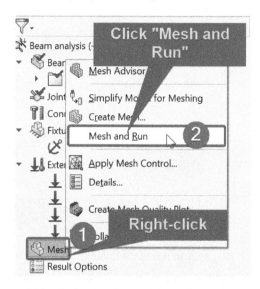

Figure 3.30 – Activating the meshing and running options

After completing *steps 1* and *2*, the study tree and graphics window will appear, as shown in *Figure 3.31a* and *Figure 3.31b*, respectively:

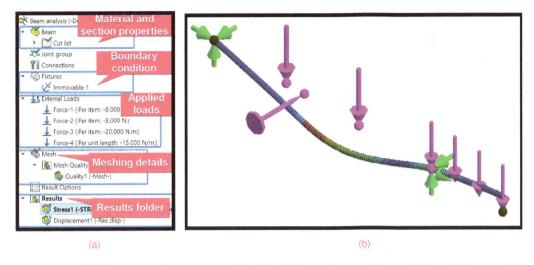

(a)   (b)

Figure 3.31 – (a) An annotated appearance of the simulation study tree after completing all the steps; (b) A partial view of the graphics area

Up to this point, we have created a sketched line of the beam and converted the sketched line into a series of structural members with volume properties using the weldment tool. Furthermore, we have specified the material property, applied fixtures, and specified three different types of loads. Besides that, we have created the mesh and run for our analysis. It is now time to conduct a post-processing analysis of the simulation results, which is what the next section is about.

Part D – Examining the results

In this section, we will address the following questions that were part of the problem statement:

- What is the maximum vertical deflection of the beam and the location of this maximum deflection?

- What is the graphical variation of the shear force and the bending moment along the length of the beam?

Let's start with the first question about obtaining the maximum deflection.

Obtaining the maximum vertical deflection of the beam

To retrieve the maximum deflection experienced by the beam, perform the following steps:

1. Right-click on **Results**, and then select **Define Displacement Plot** (*Figure 3.32*).

Figure 3.32 – Activating Displacement Plot PropertyManager

This opens up **Displacement Plot PropertyManager**, as shown in *Figure 3.33*:

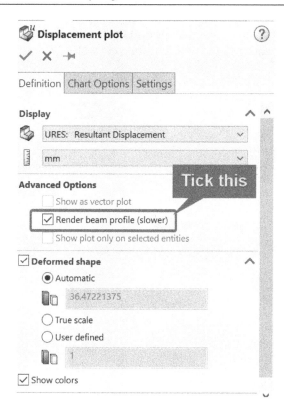

Figure 3.33 – Displacement Plot PropertyManager

You will notice that **Displacement Plot PropertyManager** has three tabs:

a. **Definition**

b. **Chart Options**

c. **Settings**

Various customizations can be done to our results using each of these tabs as follows.

Within **Displacement Plot PropertyManager**, select the following options (summarized in *Figure 3.34*):

2. Under the **Definition** tab, select **UY: Y Displacement** (*Figure 3.34a*).

3. Move to the **Chart Options** tab, under the **Display** options, and check the **Show min annotation** and **Show max annotation** boxes. While still within the **Chart Options** tab, under **Position/Format**, change the number format to floating for a more readable display of the numerical values (*Figure 3.34b*).

4.   Move to the **Settings** tab and, under **Boundary Options**, check the **Superimpose model on the deformed shape** box, and then click **OK** (*Figure 3.34c*).

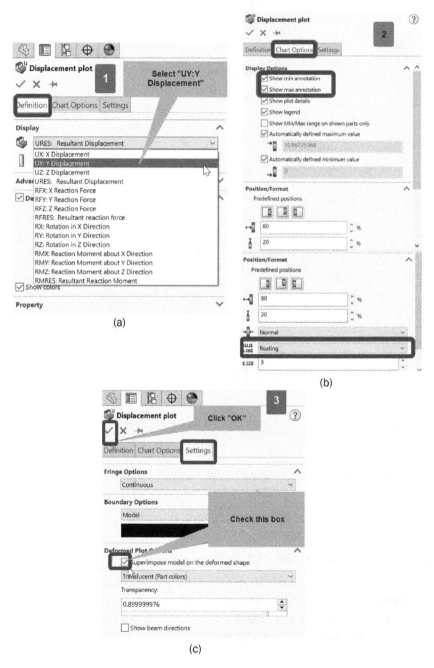

(a)

(b)

(c)

Figure 3.34 – Selecting the options for the displacement result along the Y axis

After completing *steps 1–3*, the graphic window will be updated with the displacement plot. *Figure 3.35* depicts the deflected shape of the beam (viewed in the front plane). From this figure, we can see that the beam experiences a maximum deflection value of **10.967 mm** (downward) at a point to the right of point *C*:

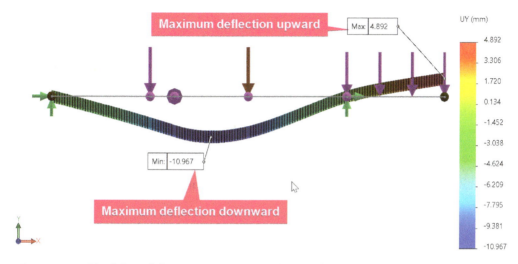

Figure 3.35 – The deflected shape superimposed on the undeflected shape of the beam (front view)

To check whether this maximum deflection value is acceptable, one rule of thumb (depending on the professional code) is to consider whether the maximum deflection value exceeds L/240, where L is the length of the beam. Going by this, since the length of our beam is L = 4 m, then L/240 = 16.67 mm. Therefore, the conclusion is that since L/240 is greater than the maximum deflection we found in the simulation, the deflection of the beam is relatively acceptable. Nonetheless, take note that the value of the maximum deflection must be used together with other stress-based failure criteria to make the right judgment about the safety of the beam.

Important Note

For a further discussion of deflection limits, you can check *Chapter 9, Simulation of Components under Thermo-Mechanical and Cyclic Loads*, in the book by *Mott* and *Untener [4]*. However, for more in-depth coverage of failure theories in the context of designing engineering structures, you can consult the books by *Collins, et al. [5]* and *Brown [6]*.

Beyond the knowledge of the maximum deflection, it is also desired to examine the visual variation of the internal resistance loads within the beam, which is what we will do next.

Obtaining the shear force and bending moment diagrams

In this subsection, we will obtain the visualization of the internal forces.

For simple beams that are loaded with just transverse forces, the internal resistant loads take the form of **shear force** and a **bending moment**. These internal resistance loads are technically responsible for the stresses that develop within beams/frames. This means that by knowing the variation of these internal resistance loads, we can graphically locate a segment of the beam/frame that will be susceptible to high stress. In practice, information about the highly stressed segment can be deployed to decide upon the segment of the beam that needs to be reinforced with more materials.

In what follows, we will start with an examination of the shear force diagram and then proceed to examine the bending moment diagram.

Shear force along the entire span of the beam

Let's generate a visualization of the shear force variation along the span of the beam:

1. Right-click on **Results**, and then select **Define Beam Diagrams** (*Figure 3.36*).

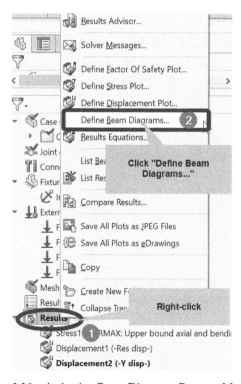

Figure 3.36 – Activating Beam Diagrams PropertyManager

Within **Beam Diagrams PropertyManager**, select the following options (summarized in *Figure 3.37*):

2.   Under the **Definition** tab, select **Shear Force in Dir1** (*Figure 3.37a*).

3.   Move to the **Chart Options** tab, under **Display Options**, and check the **Show min annotation** and **Show max annotation** boxes. While still within the **Chart Options** tab, under **Position/Format,** you should change the number format to **floating** for a more readable display of the numerical values (*Figure 3.37b*).

4.   Click **OK**.

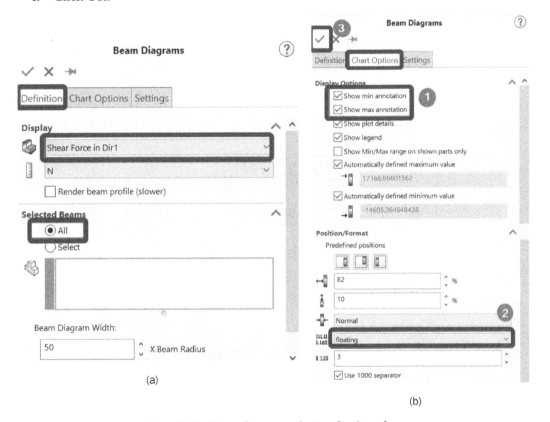

(a)

(b)

Figure 3.37 – Beam diagrams selection for shear force

5.   In the graphic window, change the view orientation to **Front View** (*Figure 3.38*):

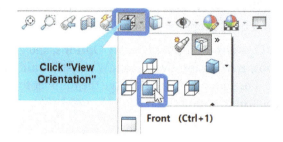

Figure 3.38 – Updating the view orientation to Front View

Once *steps 1–5* are complete, the graphic window will be updated to reveal the variation of the shear force along the length of the beam, as shown in *Figure 3.39*:

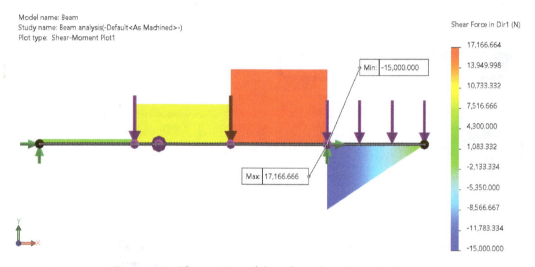

Figure 3.39 – The variation of shear force along the entire beam

The numerical values of shear force and the distribution of these values are indicated in the color legend beside the plots. As you can see from the plot, segment *EF* (carrying the distributed load) is under the effect of negative shear forces (compression), while the other segments are under the influence of positive shear forces.

In the preceding steps, we have obtained the shear force diagram for the entire length of the beam. However, it is also possible to obtain the shear force diagram for a partial segment of the beam. Indeed, in *[3]*, only a partial solution is provided for this problem. This partial solution indicates that the value of shear force immediately to the right of the 1 m position (measured from the left end) or simply to the right of segment *AB* is *9.17 kN*. Also, the value of the shear force immediately to the right of the 3 m position (measured from the left end) is reported as *-15 kN*.

To compare our simulation results with these values, we need to obtain the variation of shear force in only partial segments of the beam. This is done as follows:

1. Right-click on **Results**, and then select **Define Beam Diagrams** (similar to *Figure 3.36*).

    Within **Beams Diagram PropertyManager,** perform the following steps.

2. In the **Definition** tab, under **Display**, select **Shear Force in Dir1** (*Figure 3.40a*).

3. In the **Definition** tab, under **Selected Beams**, click on **Select**, and then pick segment 2 and segment 5 in the graphic window.

4.   Move to the **Chart Options** tab to pick the options shown in *Figure 3.40b*

5.   Click **OK**.

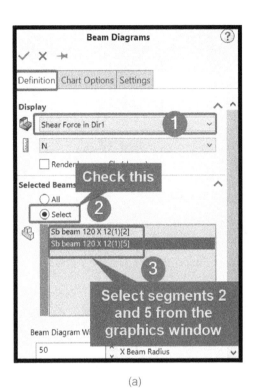

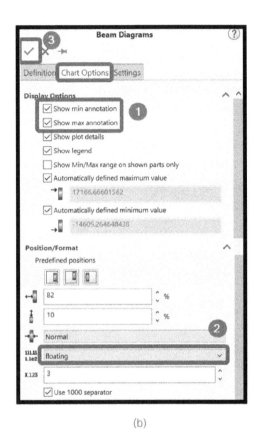

(a)   (b)

Figure 3.40 – Options for the beam diagrams concerning a partial view of the shear force diagram

6.   In the graphic window, change the view orientation to **Front View**.

With the completion of *steps 1–6*, the graphic window is updated to show the variation of shear force over segments 2 and 5 of the beam, as shown in *Figure 3.41*:

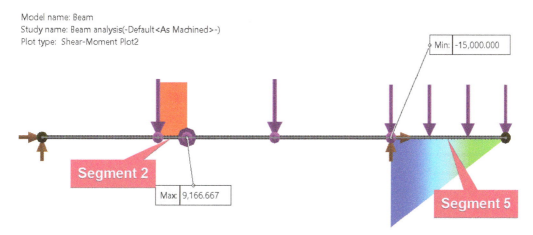

Figure 3.41 – Shear force diagram for two selected segments of the beam

From *Figure 3.41*, you can observe that the shear force to the right of the 1 m and the 3 m marks is *9166.67 N* (which approximately equals 9.17 kN) and *-15000 N* (which approximately equals 15 kN), respectively. This shows that there is a good agreement between our simulation results and that reported in *[3]*.

This now completes our evaluation of shear force and its variation along the length of the beam. Next, we move on to the graphical evaluation of the bending moment.

Obtaining the bending moment over the entire length of the beam

To visualize the variation of the bending moment along the entire span of the beam, follow these steps:

1. Right-click on **Results**, and then select **Define Beam Diagrams** (similar to *Figure 3.36*).

2. Within **Beam Diagrams PropertyManager**, in the **Definition** tab, under **Display**, select **Moment about Dir2** (*Figure 3.42a*).

3.   Move to the **Chart Options** tab to pick the options shown in *Figure 3.42b*:

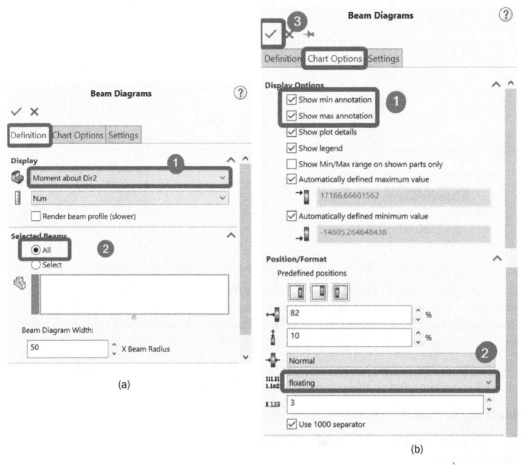

(a)

(b)

Figure 3.42 – Options for Beam Diagrams PropertyManager concerning the bending moment diagram

4. Change the view orientation to **Top View** (as shown in *Figure 3.43*).

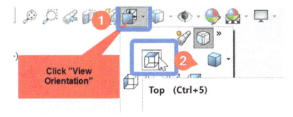

Figure 3.43 – Changing the view orientation to Top View

Figure 3.44 shows the variation of the bending moment. The figure indicates that immediately to the right of the 3 m position, the bending moment value is *7500 Nm*, which matches what was reported in *[3]*. Again, the distribution of the variation of the bending moment values is indicated in the color legend beside the plot.

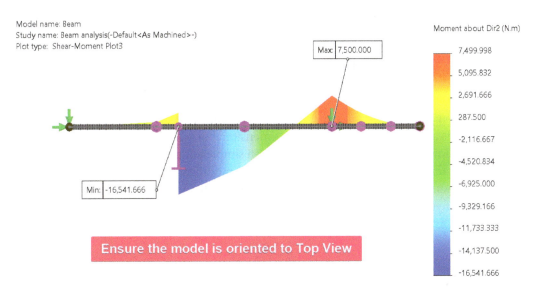

Figure 3.44 – The variation of the bending moment diagram over the entire beam

Just to reiterate, the importance of both the shear force and bending moment diagrams is to investigate the region/segment of the beam that is heavily strained. Once such a region is determined, the information can be used to reinforce the beam and hence increase its strength accordingly.

This now wraps up the solution to the questions posed for the case study. As you will have noticed, many more results can be obtained and examined depending on our objectives. As we continue our exploration in the coming chapters, we will be uncovering other important features of interest.

The concepts that we have demonstrated using the case study can be extended to many other interesting problems, as we will see in the following section.

Analysis of plane and space frames

In the case study demonstrated so far, we have deployed the beam element within the SOLIDWORKS Simulation library to simulate the structural behavior of a beam structure with different types of loads. The procedure outlined in the presented case study is perfectly suitable to study two-dimensional/plane frames. This is because a plane frame is just a collection of beam structures oriented in the two-dimensional plane. The exercise section contains a question on the extension of the beam element for the analysis of frames and a complete solution is available for download. Additionally, the procedure of this chapter is also suitable to study three-dimensional or what we generically call space frames. However, for this, we need to make a simple adjustment to *Part A* of the procedure (*creating the skeletal lines*). What does this modification involve? Basically, in the case of space frames, the skeletal lines must be situated in a 3D space, which is relatively easy if you have familiarity with SOLIDWORKS modeling. A problem on this is included in the exercise and a complete solution file is available for download.

Summary

In this chapter, we have explored the fundamental procedures involved in the static analysis of beams. Building on our knowledge of the weldment tool introduced in *Chapter 2, Analyses of Bars and Trusses*, we have demonstrated how to create the model of a beam and the transformation of the model into a beam structural member. Beyond this, the chapter entails additional knowledge that relates specifically to the simulation of beams. Some of these include the following:

- How to employ critical points along the length of the beam to create the line segments
- How to rotate a weldment profile to achieve the desired orientation

- How to apply a distributed load on a beam segment

- How to apply a moment at a point on a beam and the SOLIDWORKS Simulation convention

- How to extract the maximum deflection of a beam

- How to visualize the shear force and bending moment diagrams

You saw how to handle the simulation of trusses and beams/frames in *Chapters 2, Analyses of Bars and Trusses* and *Chapter 3, Analyses of Beams and Frames,* respectively. In the next chapter, you will learn about the analysis of components loaded with torques, and in doing so, you will familiarize yourself with how to use the **beam element** without employing the weldment tool.

Exercises

1. *Figure 3.45* shows a beam made of ASTM-A36 structural steel with a triangularly distributed load. The beam is made of S (American standard) cross-section – S120 x 12. Use SOLIDWORKS Simulation to evaluate the shear force and bending moment diagrams.

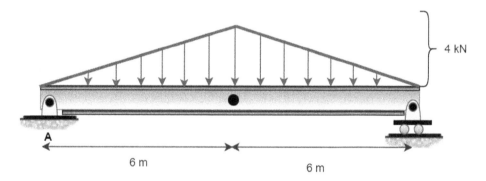

Figure 3.45

2. *Figure 3.46* shows 2D schematics of two bicycle frames that are to be compared for structural performance. The frames are made of aluminum alloy 6061 and are all 1" circular tubing (ANSI pipe 1" sch 40). Use SOLIDWORKS Simulation to determine and compare the maximum deflection and the maximum bending stress of the two designs.

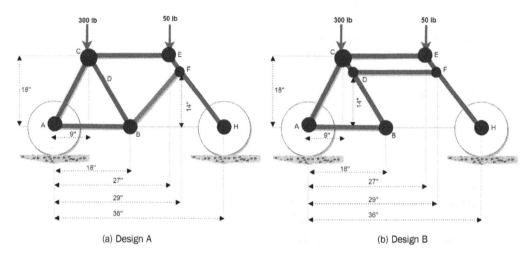

(a) Design A   (b) Design B

Figure 3.46

3. *Figure 3.47* shows a 3D space frame formed with tubular beams derived from ASTM-A36 structural steel. The beam members have the same cross-section (ISO square tube – 80 x 80 x 5 mm). Use SOLIDWORKS Simulation to do the following:

a. Determine the maximum vertical displacement of the frame.

b. Determine the maximum bending stress of the frame.

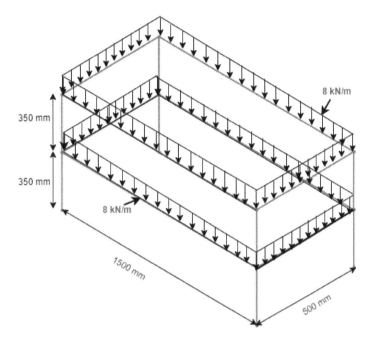

Figure 3.47

Further reading

- *[1], Structural Analysis, R. C. Hibbeler, Prentice Hall*

- *[2] Mechanics of Materials, A. Bedford and K. M. Liechti, Springer International Publishing*

- *[3] Engineering Mechanics: Statics, R. C. Hibbeler, Prentice Hall*

- *[4] Applied Strength of Materials., R. L. Mott and J. A. Untener, CRC Press, Taylor & Francis Group*

- *[5] Mechanical Design of Machine Elements and Machines: A Failure Prevention Perspective, J. A. Collins, H. R. Busby, and G. H. Staab, Wiley*

- *[6] Mark's Calculations for Machine Design, T. H. Brown, McGraw-Hill Education*

4
Analyses of Torsionally Loaded Components

One interesting aspect of SOLIDWORKS simulation is that it facilitates the analysis of components with virtually all types of loads that we meet in practical circumstances. In *Chapter 2, Analyses of Bars and Trusses*, we studied truss structures, which are made of bars (members under the effect of axial loads). In *Chapter 3, Analyses of Beams and Frames*, we shifted our attention to beams and frames (focusing only on transverse loads applied perpendicular to the axis of these structures).

One fairly basic type of load we have not studied is the torsional load. The goal of this chapter is to cover structures with this type of load. Along the way, we will demonstrate a suite of new features within the SOLIDWORKS simulation environment. Primarily, we will walk through the following topics:

- Overview of torsionally loaded members
- Some strategies for dealing with multi-segment torsionally loaded components in a SOLIDWORKS simulation
- Getting started with an analysis of torsionally loaded members in a SOLIDWORKS simulation

Technical requirements

You will need to have access to the SOLIDWORKS software with a SOLIDWORKS simulation license. You will also need to have acquired some knowledge of stress, strain, and material behavior.

You can find the sample files of the models required for this chapter here: `https://github.com/PacktPublishing/Practical-Finite-Element-Simulations-with-SOLIDWORKS-2022/tree/main/Chapter04`

Overview of torsionally loaded members

This section provides basic background information regarding torsionally loaded members. It gives the objectives of simulating these structures and their applications.

Components that are torsionally loaded may come under various categories, for instance:

- Shafts (cylindrical bars) that are under the action of twisting moments, or non-cylindrical one-dimensional structures under the action of pure twisting moments

- Beams that are loaded with shear loads for which the points of application do not coincide with the shear center of the beam section

- Grids that involve a collection of beams arranged as a framework supporting both transverse and torsional loads

- Space frames with members under the effect of torsional loads combined with other types of loads (such as axial plus bending loads)

All four categories listed above have notable applications. Nevertheless, in the rest of this chapter, we will restrict ourselves to the first category. Reasons for this include brevity and the fact that their wide applications can be easily spotted in our surroundings. Besides, they represent a straightforward way to demonstrate the concept of twisting deformation and of shear stress. Moreover, this category also represents the workhorse of rotating machinery – from the shaft of a car experiencing a torque from the engine, or a construction auger, to propeller shafts for wind and hydroelectric power generation *[1]* and *[2]* (see the *Further reading* section). Furthermore, when we use simple machines such as a lug wrench or screwdriver, torsional loading is at play.

Figure 4.1 shows some examples of machines that work based on the behavior of torsionally loaded components experiencing twisting moments.

Principally, when a shaft is torsionally loaded, shear stresses develop within it. Furthermore, the shaft experiences a twisting deformation (called the angle of twist) about its longitudinal axis. Consequently, a primary aim of the static analyses of torsionally loaded components relates to finding these quantities (that is, *shear stresses and angle of twist*) to help in design tasks.

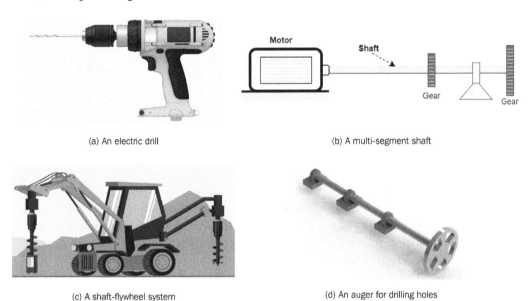

(a) An electric drill

(b) A multi-segment shaft

(c) A shaft-flywheel system

(d) An auger for drilling holes

Figure 4.1 – Some applications of torsionally loaded components in practice

The four examples shown in *Figure 4.1* are indicative of various types of torsionally loaded shafts. But there are some subtle differences in the way we can analyze them. For instance, the shaft facilitating the transmission of the load from the motor (see *Figure 4.1b*) and the one supporting the flywheel (in *Figure 4.1c*) exemplify the kind of components suitable for the approach we will demonstrate in this chapter. The other two examples are best analyzed using the approach we will present in *Chapter 6, Analysis of Components with Solid Elements*.

Having described some features of torsionally loaded shafts, we can now expand upon the strategy for their computer analysis.

Strategies for the analysis of uniform shafts

This section describes the structural details and the modeling strategy for the analysis of shafts using the SOLIDWORKS simulation environment.

Structural details

The technical details presented for the analysis of beams/frames in *Chapter 3, Analyses of Beams and Frames*, are also applicable to the analysis of simple shafts. This means you will need to know the following technical details ahead of the analysis:

- Dimensions:

  - Details of the cross-section

  - Geometric length

- Material properties

- External loads applied to the shaft

- The support provided to the shaft to prevent rigid body motion

Modeling strategies

The strategy to be adopted for creating the geometric model of a shaft to be analyzed will often depend on the complexity of the structural details. Assuming we restrict ourselves to long straight shafts with simple and uniform cross-sections, then we may choose to adopt either of the following approaches:

- Model the shaft using the method of extruded cross-sections

- Model the shaft based on the application of the **weldments** tool

When a shaft is created using either of the preceding approaches, then the simulation of its behavior can be conducted using the **beam element** within the SOLIDWORKS simulation engine.

Now, we have already dealt with the use of the weldments tool in *Chapters 2, Analyses of Bars and Trusses*, and *Chapter 3, Analyses of Beams and Frames*. Therefore, in this chapter, we will adopt the second approach so that it becomes obvious that we can, in fact, use the beam element without going through the use of the weldment tool.

But irrespective of which modeling approach you adopt, be mindful of the idea of critical positions we mentioned in *Chapter 3, Analyses of Beams and Frames*. With shafts, as with beams, the extruded sections should (in most cases) be created based on the knowledge of critical positions. To illustrate the concept of critical positions, *Figure 4.2* shows a multi-segment shaft with the critical positions labeled from *A – G*. As you will observe, the following points are treated as critical positions:

- The beginning and end of the members in a multi-segment shaft

- A point where a concentrated torque is applied

- The beginning of a distributed torque
- The end of a distributed torque
- A point where there is a change of cross-section

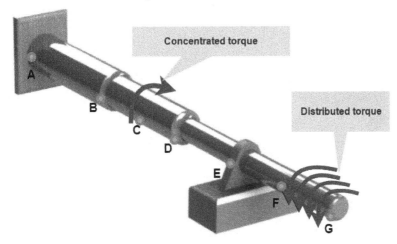

Figure 4.2 – An illustration of the critical positions (labeled A - G) in a multi-segment shaft

- A point where there is a change of material properties
- All points of support
- All points of connections

This list is by no means exhaustive, but it should give you an idea of when to use this concept. Note that it is not always necessary to treat these points as critical. For instance, if you are not interested in reading the stress or the deformation of segment BD or at points B and D (*Figure 4.2*), then you don't have to treat them as critical positions.

Characteristics of the beam element in a SOLIDWORKS simulation library

We have already used the beam element in *Chapter 3, Analyses of Beams and Frames*. Hence, you may be prompted to ask: How is it possible that we can use a beam element to analyze torsionally loaded components? Well, the answer lies in the fact that the beam element provided in the SOLIDWORKS simulation library is a *three-dimensional element that can support three translational displacements along the x, y, and z axes, and three rotational displacements about these three axes.* The rotational displacement about the longitudinal axes is what corresponds to the twisting deformation in the analysis of torsionally loaded components such as shafts. Furthermore, shear stress from torsional loads is one of the six stress components that can be extracted when the beam element is used.

With the information provided in this section, it is time to walk through a case study to demonstrate the analysis of a component with torsional loads.

Getting started with analyses of torsionally loaded members

The case study in this section illustrates the use of the beam element without relying on the weldment tool for analyses of components with torsional loads. Further, it demonstrates how to define a custom material within the SOLIDWORKS simulation environment and shows the retrieval of shear stress and angle of twist.

Time for action – Conducting a static analysis of a multi-material, stepped shaft with torsional loads

Problem statement

A compound shaft is formed by bonding an aluminum rod, AB, to the brass rod, BD. The shaft is loaded with torque at positions B and A, as shown in *Figure 4.3a*, while the length of its segment and a sectioned view of the shaft is revealed in *Figure 4.3b*. The *DC* portion of the brass segment is hollow and has an inner diameter of 40 mm. The D end of the shaft is connected to a wall, which is acting as fixed support. We are interested in simulating the loading of the shaft and consequently obtaining the following:

- The angle of twist of end A

- The maximum shear stress in the brass segment

This problem is inspired by exercise 3.38 in the book by Beer, et al. *[2]*. In the solution, the authors obtained a value of *6.02 degrees* for the angle of twist at position *A*. Can we use a SOLIDWORKS simulation to verify this answer?

Let's dig in.

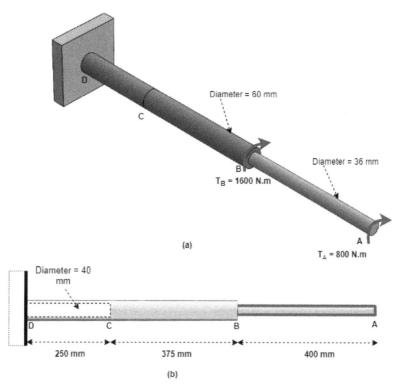

(a)

Diameter = 60 mm

Diameter = 36 mm

T_B = 1600 N.m

T_A = 800 N.m

Diameter = 40 mm

250 mm   375 mm   400 mm

(b)

Figure 4.3 – (a) Isometric view of a shaft made of brass and aluminum with torsional loads; (b) a schematic of the shaft with the length of the segments and diameter of the hollow segment

In what follows, we will use the SOLIDWORKS simulation to analyze the problem and address the questions posed. Let's start with the creation of the model, as has been the practice from the previous chapters.

Part A – Creating a 3D model of the shaft using the extrusion of cross-sections

We will create a 3D model of the shaft segments on the right plane. However, note that the model can also be created on the front plane. Before we start the creation of the 3D model of the shaft, notice that points D, C, B, and A are critical points (according to what was mentioned in the earlier sub-section – *Modeling strategies*). Specifically, points D and A are the extreme ends of the shaft, while points C and B are treated as critical points because of the change of cross-section. Of course, we also note that points B and A are the positions of the applied torques. As a rule of thumb, any segment that falls between two critical points has to be created independently, as you will see next.

To begin, start up SOLIDWORKS (**File → New → Part**) and save the file as Shaft. Ensure the unit is set to the **MMGS** unit system.

Creating segment DC

We will start with the creation of segment *DC* (the hollow segment of the shaft):

1. Click on the **Sketch** tab.

2. Click on the **Sketch** tool.

3. Choose **Right Plane**, as shown in *Figure 4.4*:

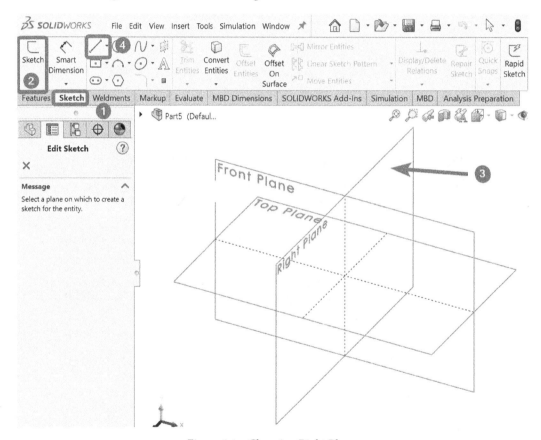

Figure 4.4 – Choosing Right Plane

4. Select the **Circle** sketching command and create two concentric circles and dimension them as shown in *Figure 4.5*.

5. Navigate to the **Features** tab, and then click on **Extruded Boss/Base**:

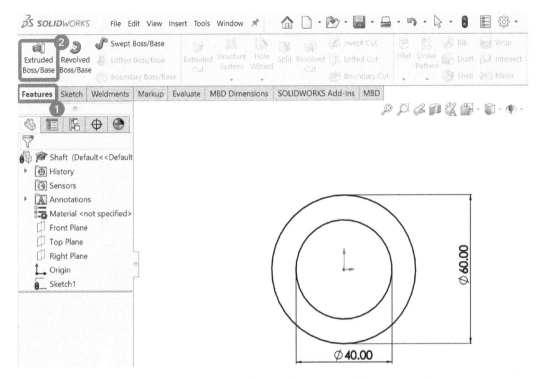

Figure 4.5 – Instantiating the extruded command (units in mm)

Step 5 brings up the **Boss-Extrude** property manager.

6. Within the **Depth** box (marked 1 in *Figure 4.6*), key in the value of **250 mm** for the depth of the extrusion.

7. Click **OK** (the green ✓) to get the extruded model of segment *DC*, as shown in *Figure 4.6*:

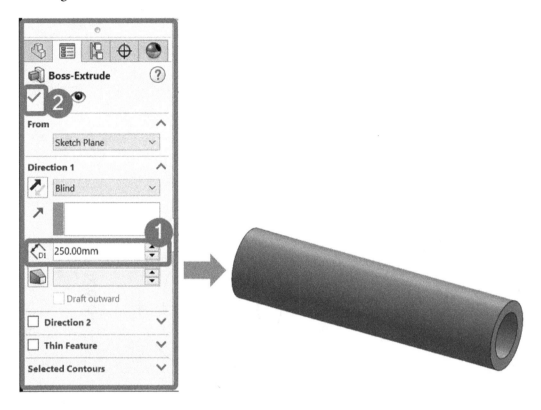

Figure 4.6 – An extruded segment of length 250 mm

We have now created the first segment, and we are set to create the second segment (*CB*) of the shaft in the next sub-section.

Creating segment CB

Segment CB is a solid segment with a diameter that is based on the outside diameter of segment *DC*. Therefore, the base sketch to be extruded for this segment will be overlaid on the right end of the completed segment *DC*. To this end, follow the steps listed next to create segment *CB*:

1. Click on the right end of segment *DC* (created in the preceding section). A context toolbar will appear shortly afterward, as shown in *Figure 4.7*.

2. Choose **Normal to** so that the face is oriented.

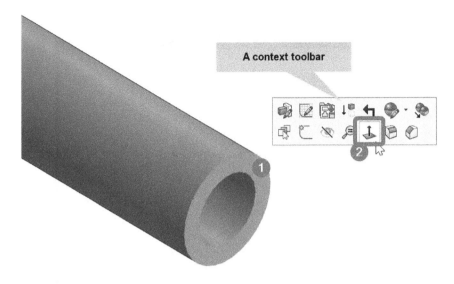

Figure 4.7 – Selections to orient the right end of segment DC

3. After the face is oriented, click on the face again so that the context toolbar re-appears.

4. Then, select **Sketch,** as shown in *Figure 4.8*:

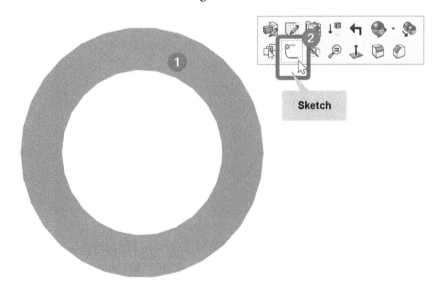

Figure 4.8 – Preparing to sketch on the oriented face

5. Pick the **Circle** sketching command to create a *60 mm* diameter circle.

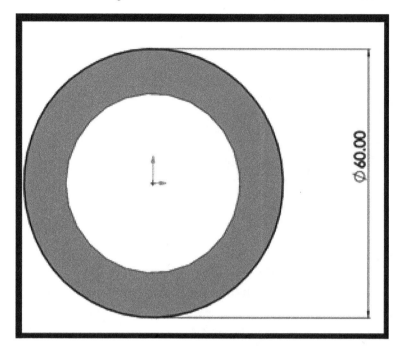

Figure 4.9 – A sketched circle with a diameter of 60 mm

6. Navigate to the **Features** tab.

7. Click on **Extruded Boss/Base** (*Figure 4.10*):

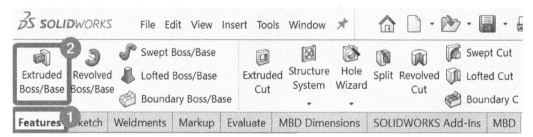

Figure 4.10 – Activating the extruding feature

The completion of *step* 7 will activate the **Boss-Extrude PropertyManager**. Now we will work with the options highlighted in the next steps that follow.

8. Under the **Depth** box (marked 1 in *Figure 4.11a*), key in the value of 375 mm for the depth of the extrusion.

9. Untick **Merge result** (marked 2 in *Figure 4.11a*).

> **Important Note**
> *Step 9* is very important. Ensure you untick **Merge result** before clicking **OK**.

10. Click **OK** (the green ✓).

Following completion of the preceding steps, you will get the extruded model of segment *CB*, as shown in *Figure 4.11b*:

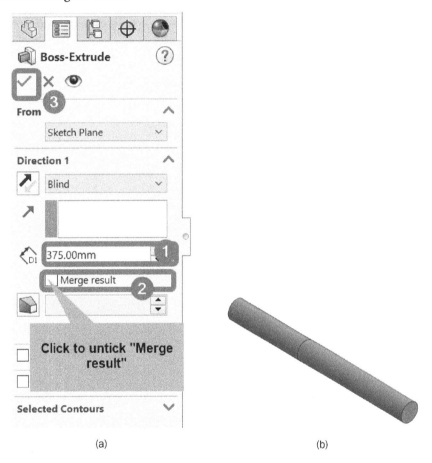

(a)   (b)

Figure 4.11 – Options for creating the second extruded segment (CB)

Note that step if *Step 9* is omitted, SOLIDWORKS will merge the two segments and a joint will not be created between the two different cross-sections. The consequence of this is that you will not be able to apply a torque or easily read the angle of twist at the intersection between the hollow and solid brass segments.

With these two segments created (that is, *DC* and *CB*), we next move on to the creation of the last segment of the shaft in the next sub-section.

Creating segment BA

Segment *BA* is also a solid segment. As we did in the preceding sub-section, we will overlay the base sketch of this segment on the right end of the just-completed segment *CB*. The steps to create segment *BA* are similar to what was outlined for the creation of segment *CB*.

To this end, follow the steps given next:

1. Click on the right end of segment *CB* (created in the preceding section). A context toolbar will appear shortly afterward.

2. Choose **Normal to** so that the face is oriented.

3. After the face is oriented, click again on the face so that the context toolbar re-appears.

4. Select **Sketch**, as shown in *Figure 4.12*:

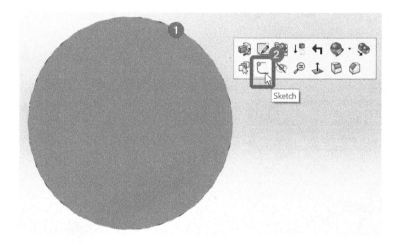

Figure 4.12 – Preparation for sketching on the oriented face of segment CB

5. Pick the **Circle** sketching command to create a **36 mm** diameter circle.

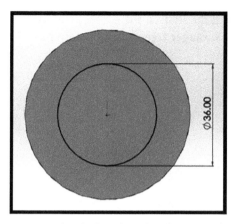

Figure 4.13 – A sketched circle with a diameter of 36 mm

6. Navigate to the **Features** tab, and then click on **Extruded Boss/Base**.

    Again, this step depicts the **Boss-Extrude** property manager (see *Figure 4.14*).

7. In the **Depth** box, key in the value of 400 mm for the depth of the extrusion.

8. Untick the **Merge result** checkbox.

9. Click **OK** (the green ✓) to get the extruded model of segment *BA*:

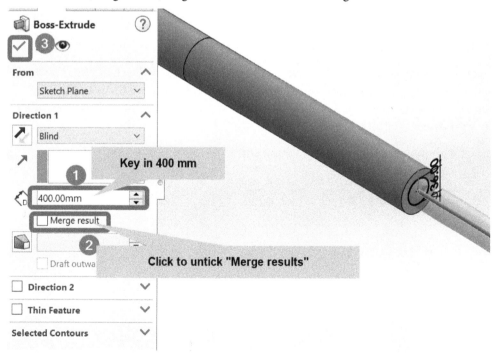

Figure 4.14 – Options for creating the second extruded segment (CB)

The new status of the model of the shaft in the graphic window should look like *Figure 4.15*. Notice that the **FeatureManager** tree should contain three bodies, as indicated:

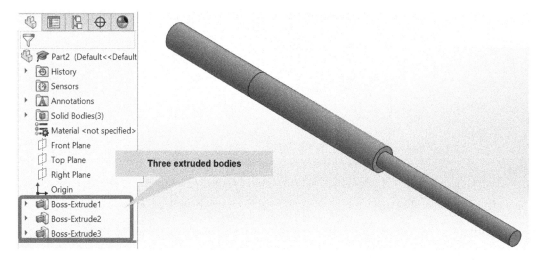

Figure 4.15 – A complete extruded model of the shaft with the distinct segments

At this point, we have completed the creation of the model of the shaft. In the next sub-section, we will launch the simulation study.

Part B – Creating the simulation study

This section comprises three sub-sections. The first sub-section will see us going through the activation steps for the simulation study, after which we shift our focus to the conversion of the extruded solid bodies to the beam element. The final sub-section deals briefly with contact settings.

Let's start with the creation of the simulation study.

Activating the Simulation tab and creating a new study

To activate the simulation commands, follow the steps highlighted next:

1. Click on **SOLIDWORKS Add-Ins**.

2. Click on **SOLIDWORKS Simulation** to activate the **Simulation** tab (*Figure 4.16*):

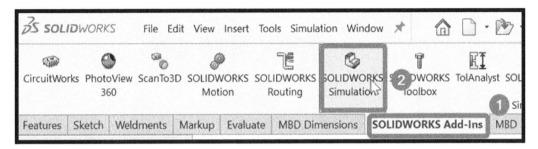

Figure 4.16 – Activating the Simulation tab

3. With the simulation tab active, click on **New Study**:

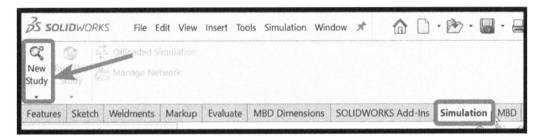

Figure 4.17 – Creating a new study

Now we will work with the options within the **Study** property manager (*Figure 4.18*).

4. Input a study name within the **Name** box, for example, Shaft analysis.

5. Keep the **Static** analysis option (selected by default).

6.   Click **OK** (that is, the green checkmark).

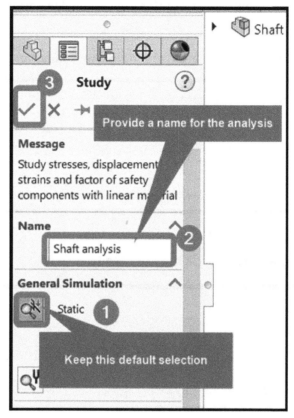

Figure 4.18 – Study names and options

With the completion of *steps* 1-6, the simulation study tree becomes active, as depicted in *Figure 4.19*:

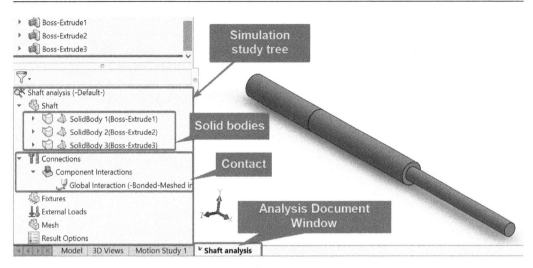

Figure 4.19 – Simulation property manager

Two details can be observed if you examine the simulation study manager (*Figure 4.19*):

- The first observation is that each extruded segment of the shaft has been converted into a solid body. Because we have three segments, there are three solid bodies (in other words, **SolidBody1**, **SolidBody2**, and **SolidBody3**). This differs from what happened when we used the weldment approach (in *Chapter 2*, *Analyses of Bars and Trusses*, and *Chapter 3*, *Analyses of Beams and Frames*), where the parts are automatically converted to beam elements after initiating the study.

- The second observation relates to the fact that SOLIDWORKS has defined a **Global Interaction** under the **Connections** property name. This global interaction describes the contact defined between each of the three segments. Each of these two observations is expanded upon in the next two sub-sections.

Let's start with the transformation of the extruded solid bodies.

Note – Change of Terminology in SOLIDWORKS 2021-2022

If you are using an earlier version of the SOLIDWORKS simulation, what you will see under the **Connections** property name is **Component Contacts** and **Global Contact** instead of **Component Interactions** and **Global Interaction**.

Converting the extruded solid bodies to beams

Since we wish to use the beam element, we will have to convert the solid bodies created under the simulation study tree into beams by following the next two steps:

1. Select all three bodies as shown in *Figure 4.20* (for example, select **Solidbody1**, press and hold down the control key, and then select the other two bodies).

2. Right-click within the region containing the selected bodies and then click on **Treat selected bodies as beams**.

> **Note**
>
> If you forget to convert the extruded solid bodies into beams, then SOLIDWORKS will automatically employ solid elements for the simulation. Practically, there is no harm in using solid elements to analyze the shaft. However, if that happens, you will not be able to easily extract the angle of twist for the shaft. Why? As you will see in *Chapter 6, Analysis of Components with Solid Elements*, solid elements do not support rotational degrees of freedom.

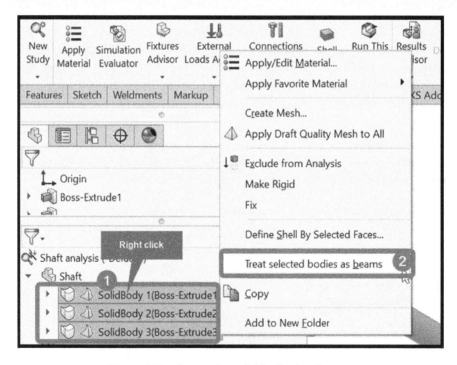

Figure 4.20 – Converting solid bodies into beams

After completing *steps* 1-2, there will be a change in the symbol attached to each of the three bodies from ▽⟳ to 🦴 (which symbolizes beams). Along with this change, joints will be created at the critical positions of the shafts, as shown in *Figure 4.21*:

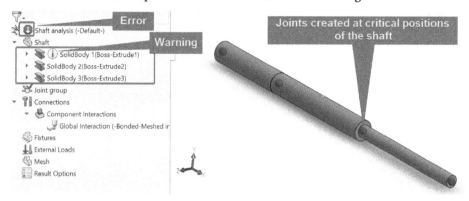

Figure 4.21 – Creation of beams and joints

Meanwhile, in *Figure 4.21*, you will notice the warning sign attached to the first body. As a result of this warning, an **Error** notification appears beside the analysis name. The beam warning arises from the fact that the length to diameter ratio for the first segment is less than *10*, which is true. However, the consequence of violating this warning is not fatal for this analysis and can be safely ignored (as will be confirmed by our final results).

In many instances, when an error notification is attached to our analysis name, it is a good idea to find out the reason. This can be done as illustrated in *Figure 4.22*. Usually, after clicking **What's wrong?**, a diagnostic message about the cause of this error will appear.

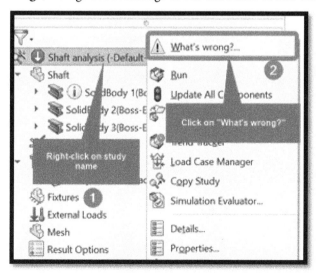

Figure 4.22 – Finding out more about an error/warning

Information

Beam Warning: Hovering over the beam with the warning sign (*Figure 4.21*) will give information that this segment is too short to be considered a beam. This is because the classical beam theory demands that the slenderness ratio (that is, the ratio of the length/characteristics dimension, for example, L/width or L/height or L/diameter) of any segment to be considered a beam should be greater than *10*.

So far, you have covered the steps for the transformation of extruded bodies into beams. You have also taken a short peek into the evaluation of an error warning within the simulation study.

Let's wrap up this sub-section by checking on the contact settings.

Scrutinizing the interaction settings

You will recall that during the modeling phase in section A, each segment of the shaft segment was extruded *without being merged* with the preceding one. This is an important strategy for dealing with multi-segment components. However, in realistic engineering scenarios, the segments are to be perfectly bonded. Based on this logic, a SOLIDWORKS simulation automatically defines interactions between the unmerged extruded bodies. The question that arises from this is: What type of interaction is defined between the bodies?

To answer the above question, this sub-section briefly discusses interaction settings generated by a SOLIDWORKS simulation between the three segments of the shaft.

To this end, perform the following steps to scrutinize the interaction settings:

1. Expand each of the beams to expose the contact setting between each segment, as shown in *Figure 4.23*:

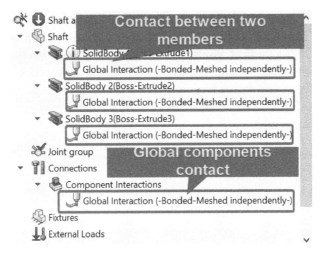

Figure 4.23 – Appearance of Component interaction settings

2. Next, right-click on one of the exposed **Global Interaction** option and choose **Edit Definitions**. This will lead to the appearance of the **Component Interaction** property manager. Keep the options as shown in *Figure 4.24*:

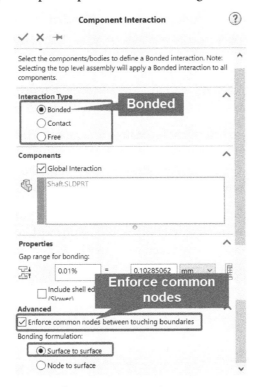

Figure 4.24 – Component interaction property manager

You will notice that we have a **Global Interaction** defined under the **Connections** property name. The **Global Interaction** defines the contact condition that applies to all components in multi-body parts or assemblies. It is this definition that is propagated under the shaft segments as shown in *Figure 4.23*. Principally, interaction settings define the relationship between entities that have initial contact or have the possibility to touch each other during loading. As you will notice in *Figure 4.24*, three types of interactions are listed under **Interaction Type**:

- **Bonded**
- **Contact**
- **Free**

> **Note**
>
> If you are using an earlier version of the SOLIDWORKS simulation, what you will see within the Component Interaction property manager under the **Interaction Type** option are: **No Penetration** (this is now known as **Contact**); **Bonded**; and **Allow Penetration** (this is now known as **Free**).

Now, as seen in *Figure 4.24*, the default selections are **Bonded** with the option to **Enforce common nodes between touching boundaries**. So, what does this mean? In effect, this choice indicates that the components affected by this choice will act as if they are welded during simulation. As a result, the **Bonded** interaction type is perfect for this simulation since we will want the segments of the shaft to be perfectly in touch with each other to account for the artificial joints we introduced at the critical positions.

Meanwhile, the simulation environment may show other categories of interaction (not seen in *Figure 4.24*) depending on the types of bodies that are in interacting with one another. For this reason, in *Chapter 7, Analyses of Components with Mixed Elements*, we will re-visit the options for interactions when we combine different higher-order elements.

At this point, we have completed the initiation of the simulation study. We have also learned about the conversion of extruded bodies into beams without making use of the weldments tool. Furthermore, we have taken a cursory look at interaction settings.

Next, we will learn about how to create a custom material and apply material properties to different segments of the shaft.

Part C – Creating custom material and specifying material properties

In *Chapter 2, Analyses of Bars and Trusses*, and *Chapter 3, Analyses of Beams and Frames*, we have applied a single type of material to the components that we analyzed. Moreover, in those chapters, we accepted the values of the material properties contained in the SOLIDWORKS materials database.

In this section, you will learn how to create a custom material from a base material within the library.

You may be wondering why we need to create a custom material. Recall that from the problem statement, segments *DC* and *CB* are made of brass (*G = 39 GPa*), while segment BA is made of aluminum alloy (*G = 27 GPa*). There is an aluminum alloy in SOLIDWORKS materials database with the exact shear modulus (that is, *G = 27 GPa*) as we have in the problem statement. Hence, it is a straightforward matter to apply the property for segment *BA*. However, as will be shown next, the grade of brass within the library has a shear modulus, *G = 37 GPa*, which is slightly lower than what was specified in the problem statement. Although the difference is small, we will use this opportunity to showcase how to create a custom material.

Let's get going.

Creating and applying a custom material property (custom brass)

There are a few ways to create custom materials in SOLIDWORKS. You may create a new custom material from scratch if it is not available within the materials database. But you can also work from a material that is already within the database, in which case we simply modify the values of the properties to suit our needs. The method we will demonstrate in this section adopts the latter approach.

To this end, perform the following actions under the simulation study tree:

1. Select segments *DC* and *CB* (representing the brass segments).
2. Right-click within the colored area of the selected items.

3.   Click **Apply/Edit Material...** (*Figure 4.25*):

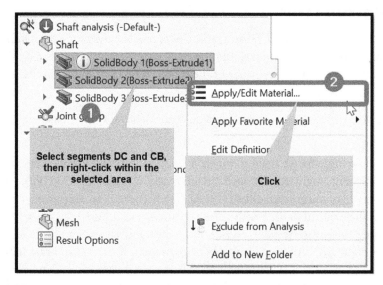

Figure 4.25 – Activating the material database

As soon as you complete *steps* 1-3, the material database appears (which you saw in *Chapter 2, Analyses of Bars and Trusses*, and *Chapter 3, Analyses of Beams and Frames*).

We are interested in brass, which is a copper alloy. Therefore, to create a custom brass from this in-built brass material, follow the steps given next.

While the material database is still on, perform the following steps:

1.   Expand the `Copper Alloys` folder to locate `Brass`.

2.   Right-click on `Brass` and choose **Copy** (*Figure 4.26*):

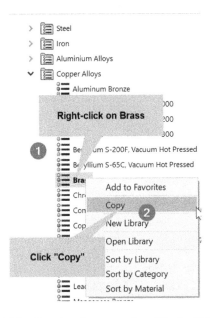

Figure 4.26 – Copying the properties of the built-in brass

3. Scroll down to the material folder named Custom Materials, right-click, and choose New Category (*Figure 4.27*):

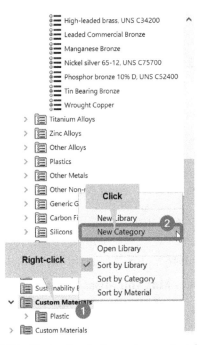

Figure 4.27 – Navigating to the custom material section

Under the Custom Materials folder, a new material folder will be created with a default name New Category.

4. Change the name to MyBrass (*Figure 4.28*).

5. Right-click on the newly created MyBrass folder and select **Paste** from the context menu that appears (*Figure 4.28*):

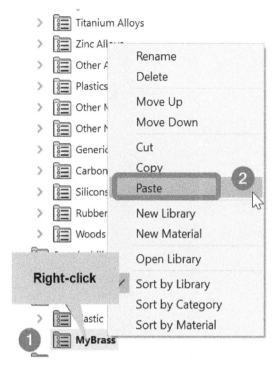

Figure 4.28 – Selection of options to create a custom brass from the base brass material

At this point, we have copied and pasted all the properties of the in-built brass. We are now ready to update the property of the brass to suit our needs. Note that after the preceding step (in other words, *step 5* above), a material named Brass is created under the MyBrass folder (see *Figure 4.29*). To avoid confusion, perform the following steps:

1. Right-click on the new Brass material and rename it Brass-39 as depicted in *Figure 4.29*.

2. Change the shear modulus to 3.9e+10, as shown in *Figure 4.29*.

3. Keep the failure criterion as **Max von Mises Stress**.

4. Click **Save**.

5. Click **Apply** and **Close** (in that order).

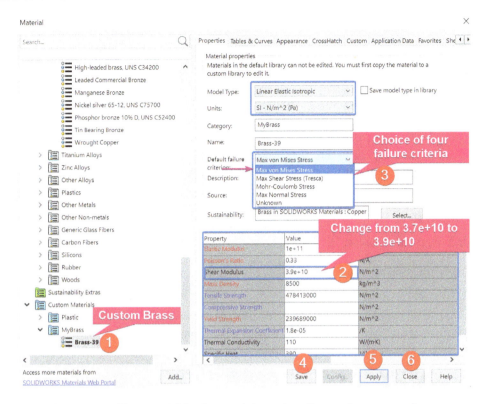

Figure 4.29 – The material database and the options for creating a custom brass

Before we transition to the next action, take note of the *four failure theories* that SOLIDWORKS supports for linear elastic isotropic materials. The names of these theories, as highlighted in *Figure 4.29*, are as follows:

- **Max von Mises Stress**
- **Max Shear Stress (Tresca)**
- **Mohr-Coulomb Stress**
- **Max Normal Stress**

Each of these criteria specifies the condition under which an isotropic material fails. The first two apply to ductile materials (such as steel). In contrast, the last two are used for the failure prediction of brittle materials. SOLIDWORKS will, by default, use the von Mises failure theory for analyses of ductile materials. The fundamental idea of this theory dictates that in a component made of ductile material, if the maximum von Mises stress exceeds the yield strength of the material from which the component is made, then yielding of the component will happen. For further details about the aforementioned theories, you may refer to the books by Beer, et al. *[2]* or Collins, et al. *[3]*, among others.

In any case, after completing *steps* 1-5, a tick mark will appear on segments *DC* and *CB*. But there are three segments within the shaft, and we need to specify the material for the last segment. This is what we do in the next sub-section.

Adding a material property (aluminum alloy)

Here, we will apply the material details to segment *BA* of the shaft. This segment is made of aluminum alloy. To apply the aluminum material property, follow the steps given next:

1. Right-click on segment *BA* and choose **Apply/Edit Material**.

2. Expand the `Aluminium Alloys` folder.

3. Click on **1060 Alloy**.

4. Click **Apply** and **Close**.

After closing the material database window, a green tick mark (✓) will appear on the part's name.

At this point, we have completed the specification of the material properties for all segments of the shaft. Next, we will walk through the application of fixtures and torques.

Part D – Applying torque and fixtures

This sub-section focuses on how to set up the application of the torques (or twisting moments) at joints *B* and *A* of the shaft. We will also carry out the application of fixtures at joint *D*.

Let's begin with the application of fixtures.

Applying a fixture at joint D

We only have a single point of support for this problem located at joint *D*. To apply the fixture to this joint, follow the steps given next.

Under the simulation study tree, perform the following actions:

1. Right-click on **Fixtures**, and then pick **Fixed Geometry** from the context menu that appears.

2. Within the **Fixture** property manager that appears, select **Fixed Geometry** (*Figure 4.30*).

3. Navigate to the graphics window and pick joint **D**.

4. Click **OK**.

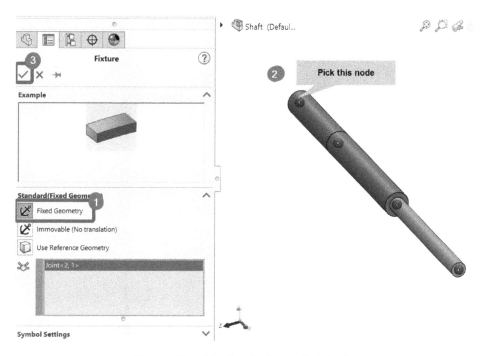

Figure 4.30 – Selecting the fixture for joint D

At the end of *steps 1-4*, the arrow symbolizing the fixed fixture will appear at joint *D*. This fixture restricts joint *D* so that all six degrees of freedom are zero.

With the fixture application completed, we now shift our attention to the application of the two twisting moments representing the payload torques.

Applying torque at joints B and A

In this sub-section, the torque we wish to apply is a type of moment that leads to the twisting of the longitudinal axis of the shaft.

Let's follow the steps given next to apply the torques at joints *B* and *A*.

Under the simulation study tree, perform the following actions:

1. Right-click on **External Loads** and select **Torque** (as shown in *Figure 4.31*):

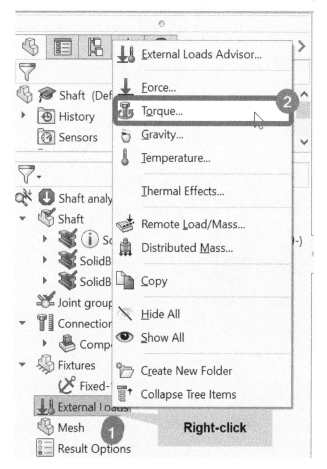

Figure 4.31 – Initiating the application of torque

Now we will work with the options within the **Force/Torque** property manager that appears.

2. Under **Selection**, click on the joint symbol (labeled 1 in *Figure 4.32*).

3. Navigate to the graphic window to click on joint **B**.

4. Under the reference box direction (labeled 3 in *Figure 4.32*), choose the **Front Plane**.

5. Under **Moment**, activate the **Along Plane Dir 1** box and input 1600 (ensure that the unit system remains an SI unit).

6. Click **OK**.

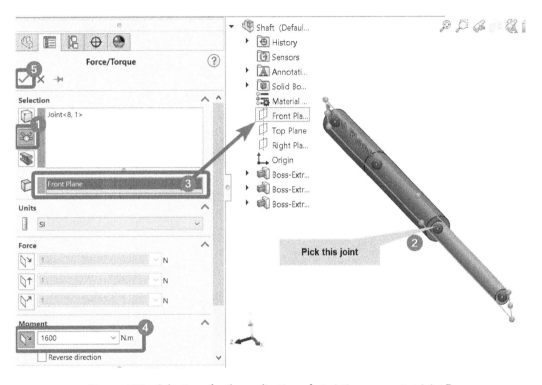

Figure 4.32 – Selections for the application of a twisting moment at joint B

Once *steps* 1-6 have been completed, a twisting moment will be applied to joint *B*.

Next, we need to apply another twisting moment to joint *A*. The steps are similar to what we have just done for joint *B*. For brevity's sake, *Figure 4.33* captured the options to be made within the **Force/Torque** window:

Figure 4.33 – Selections for the application of a twisting moment at joint A

Following the application of twisting moments at joints *B* and *A*, the simulation study tree and the model in the graphics window will likely appear as shown in *Figure 4.34*.

You will notice that the arrow representing the two torques at these joints does not appear in the graphics window (as confirmed in *Figure 4.34*). This is because the torques are directed along the length of the shaft and they have been overshadowed by the third component.

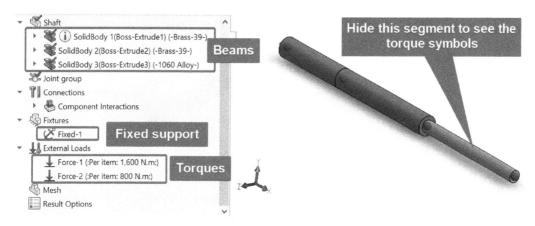

Figure 4.34 – Appearance of the simulation study tree and the model

To see that we have truly applied the torques, we will reveal the features symbolizing them by following the steps given next:

1. Navigate to **FeatureManager**.

2. Right-click on **Boss Extrude 3** (indicated by 1 in *Figure 4.35a*) and then select **Hide** from the context menu as shown in *Figure 4.35a*, the symbols representing the torques will appear as shown in *Figure 4.35b*.

3. Once you have seen the symbols, you may unhide the segment before running the analysis.

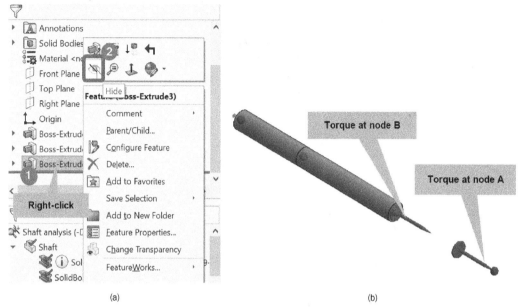

(a)   (b)

Figure 4.35 – (a) Options to hide segment BA; (b) exposing the torque symbol at joints B and A

This now concludes the application of torque on the joints of the shaft. We are getting close to completing the pre-processing tasks for the simulation.

So far, we have learned how to convert extruded entities into beams without using the weldments tool. In addition, we have covered how to create a custom material and showed how to modify the values of specific material properties. Besides, we have briefly examined contact settings and we have explored the application of torques. Now, we can move to the meshing task and running of the analysis, which is what happens in the next sub-section.

Mesh and Run

We have already examined the options to mesh and run our analysis in *Chapter 2, Analyses of Bars and Trusses*, and *Chapter 3, Analyses of Beams and Frames*. As explained in those chapters, any time we employ either the **truss** or beam element, a straightforward meshing command is **Mesh and Run**. Therefore, we will employ this same command here.

To that end, follow the steps given next:

1. Right-click on **Mesh** within the simulation study tree.

2. Select **Mesh and Run**.

After completing these two steps, the study tree will be updated with default results (**Stress 1** and **Displacement1**), and the graphic window will display the deformed shape that corresponds to **Stress1**. However, these two results are not our focus.

In the next section, we will retrieve the results that we want.

Part E – Post-processing of results

Now that we have completed the steps in the previous sections, we will now attempt to address the following questions:

1. What is the angle of twist generated at joint *A* due to the combined effects of the two payload torques?

2. What is the maximum shear stress that developed within the shaft?

Let's begin with the retrieval of the angle of twist.

Obtaining the angle of twist at joint A

As with other types of loads, when a component is torsionally loaded with a torque (also known as a twisting moment), it experiences deformation, strain, and stress. And because a twisting moment creates a rotation of the material points on the cross-section of a loaded component, one of the notable deformations that arise from it is the *angle of twist*. An angle of twist is simply a rotational displacement of one end of a loaded component with respect to another. It is often quantified with respect to the axis through which the twisting moment acts.

To obtain the angle of twist experienced by the shaft, we follow these steps:

1. Right-click on **Results** and then select **Define Displacement Plot**.

    Now we will work with the options within the **Displacement plot** property manager (summarized in *Figure 4.36*) that appears.

2. On the **Definition** tab, navigate to **Display**.

3. Select **RX: Rotation in X Direction** and select the other options highlighted in *Figure 4.36a*.

4. Move to the **Chart Options** tab, under the **Display** options, and check the **Show max annotation** box.

5. Still within the **Chart Options** tab, under **Position/Format**, change the number format to **floating**, as depicted in *Figure 4.36b*.

6.   Click **OK**.

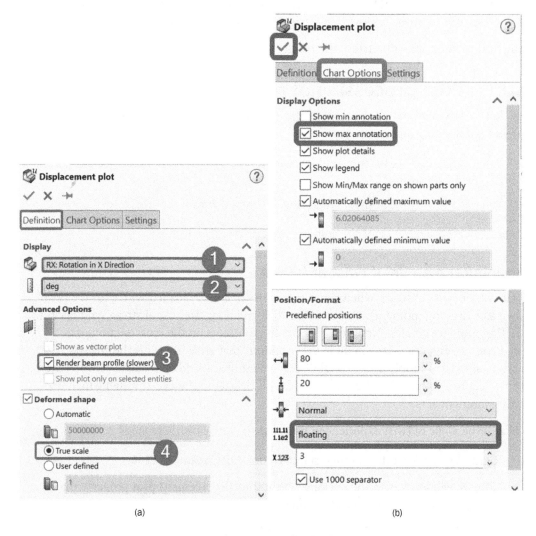

(a)   (b)

Figure 4.36 – Specifying options for the displacement plot

Figure 4.37 depicts the angle of twist (or the rotational displacement about the x axis) of joint *A*. This figure indicates that joint *A* experiences a rotational displacement of *6.02* degrees with respect to the fixed end (joint D). Furthermore, as a confirmation of the accuracy of this simulation, it is important to observe that the value of the angle of twist that we have obtained matches exactly the value reported in *[2]*.

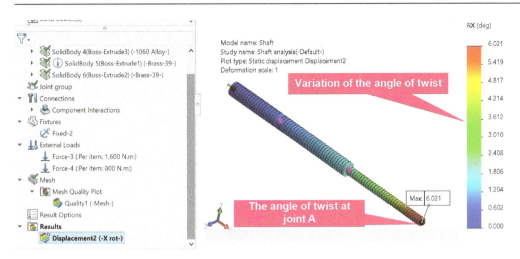

Figure 4.37 – The display of the rotational angle of twist

In *Figure 4.37*, notice that segment *DC* is bluish. This is because joint *D*, which is the end of the shaft, is fixed (that is, not allowed to rotate or move) when the torques are applied at joints *B* and *A*. Hence, the angle of twist at joint *D* is *0*. The angle twist can be seen to increase from joint *D* (blue color) to joint *A* (red color). It is worth reiterating here that the main benefit of identifying the critical positions in our components during modeling is that we can obtain the deformation data at those positions during post-processing. So, go ahead and check the angle of twists at joints *B* and *C*.

Now, beyond the deformation data, it is always a good idea to check the stress experienced by the component. Therefore, we will set up the option for the evaluation of the shear stress in the shaft. This is what happens next.

Obtaining the shear stress

In *Chapter 2*, *Analyses of Bars and Trusses*, and *Chapter 3*, *Analyses of Beams and Frames*, the major stress that develops in the components we simulated is normal stress. However, for torsionally loaded components such as the shaft we are analyzing in this chapter, the type of stress that follows the torsional load is called shear stress. As you will know, in the most general case of loading, three types of normal stress and three types of shear stress are experienced at a point within a component. To distinguish between the three types of shear stress, a SOLIDWORKS simulation refers to the shear stress caused by a torsional load as *torsional shear stress*.

To obtain the torsional shear stress, follow the steps given next:

1. Right-click on **Results** and then select **Define Stress Plot**.

    Now we will work with the options within the **Stress plot** property manager (summarized in *Figure 4.38*) that appears.

2. On the **Definition** tab, navigate to **Display**.

3. Change the unit to **N/mm^2 (MPa)** and select the other options highlighted in *Figure 4.38a*.

4. Move to the **Chart Options** tab and make the selections as depicted in *Figure 4.38b*.

5. Click **OK**.

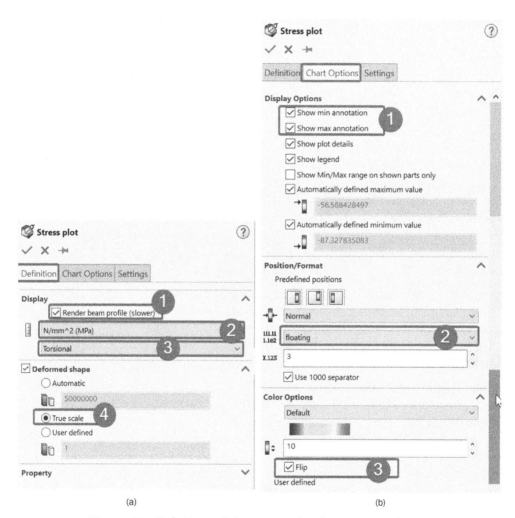

(a)   (b)

Figure 4.38 – Definition and chart options for plotting torsional stress

Figure 4.39 suggests that the absolute maximum shear stress is felt by segment BA, which is composed of aluminum alloy.

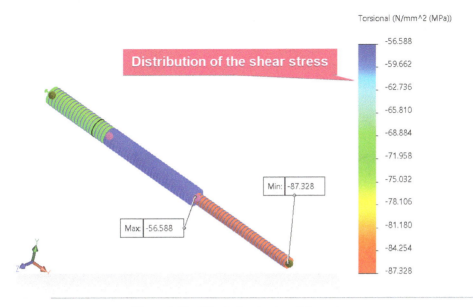

Figure 4.39 – Displaying torsional shear stress

Besides, the negative value of the stress indicates that the segment is under a state of compressive shear stress. In practice, we will want to know whether this stress will cause the component to yield. We may also wish to know the minimum factor of safety within the analyzed component. All of these could be explored further, but our primary objective is simply to demonstrate how to obtain the torsional shear stress value here.

We have now completed the simulation and obtained the answers to the questions posed for the case study of this chapter. The concepts that we have demonstrated using the case study can be extended to many other interesting problems. Some of these problems are highlighted in the next section.

Analysis of components under combined loads

In the example demonstrated so far, we have deployed the beam element for the simulation of a shaft under the influence of torsional loads. Combined with the knowledge of the last two chapters, the procedure we have outlined in the presented example can be easily extended to other problems. For instance, it is a straightforward matter to extend this to components under combined axial plus torsional loads or beams under transverse plus torsional loads. One of the questions in the exercise, inspired by the practical problem proposed in *[4]*, relates to the case of a shaft under combined loads. Furthermore, the approaches we employed in *Chapter 3*, *Analyses of Beams and Frames,* and this chapter can also be combined to study grids featuring transverse and torsional loads.

Summary

At the heart of this chapter is a case study that facilitates the exploration of the stress and deformation that stems from the application of torsional loads to an engineering component. Via this case study, we have demonstrated strategies for the following:

- Creating a multi-segment extruded component for simulation purposes
- Creating a custom material from a specific base material
- Applying torsional loads or twisting moments to a component
- Extracting the angle of twist that follows the application of torsional load

With the concepts covered in this chapter (as well as *Chapter 2, Analyses of Bars and Trusses*, and *Chapter 3, Analyses of Beams and Frames*), you are now equipped with a few fundamental concepts that can be used to analyze components that can be idealized as one-dimensional members or a collection of one-dimensional members. In the next chapter, we will unveil SOLIDWORKS capability for the analysis of axisymmetric components, which is more advanced than what we have seen up to now.

Exercise

1. *Figure 4.40* shows a three-segment shaft used to transmit the two payload torques applied at joints B and C. The shaft is composed of AISI 1010 steel with a shear modulus of 80 GPa. The length L_{AB} = L_{CD} = 200 mm, while L_{BC} = 240 mm. Likewise, segments AB and CD are of diameter 25 mm, while segment BC has a diameter of 50 mm. Use a SOLIDWORKS simulation to evaluate the maximum shear stress in the shaft and then determine the angle of twist of joint C.

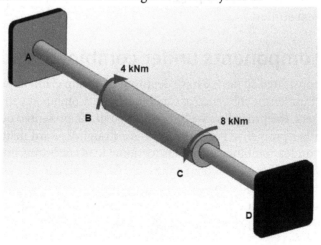

Figure 4.40 – A stepped shaft with payload torques

2. *Figure 4.41* shows a shaft acting as part of the rotating component of a helicopter. The shaft is under the effect of a combined axial and torsional load, as indicated. The material of the shaft is aluminum alloy 6061, while its cross-section is made of a tube with internal and external diameters of 34 mm and 80 mm, respectively. Use a SOLIDWORKS simulation to determine the maximum shear stress and the maximum von Mises stress.

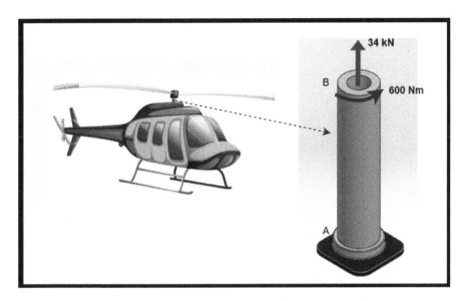

Figure 4.41 – A shaft with a combined axial and torsional loads

References

- [1] *Mechanics of Materials, Cengage Learning, B. J. Goodno and J. M. Gere, 2016.*

- [2] *Mechanics of Materials, McGraw-Hill Education, F. Beer, E. R. Jr. Johnston, J. DeWolf, and D. Mazurek, 2011.*

- [3] *Mechanical Design of Machine Elements and Machines: A Failure Prevention Perspective, J. A. Collins, H. R. Busby, and G. H. Staab, Wiley, 2009.*

- [4] *Mechanics of Materials, A. Bedford and K. M. Liechti, Springer International Publishing, 2019.*

Section 2: SOLIDWORKS Simulation with Shell and Solid Elements

Having covered the basic elements in the first section of the book, the next three chapters sketch out the application of SOLIDWORKS Simulation for analyzing components with more complicated geometries than those treated in the previous section. By the end of this section, you will have become familiar with the Simulation workflow involving shell and solid elements and the various options that are available when analyses are conducted with these advanced elements.

This section comprises the following chapters:

- *Chapter 5, Analyses of Axisymmetric Bodies*
- *Chapter 6, Analyses of Components with Solid Elements*
- *Chapter 7, Analyses of Components with Mixed Elements*

5
Analyses of Axisymmetric Bodies

In the previous three chapters, we dealt with the analyses of components that are analyzed using either a **truss** or **beam element**. Using these elements, we have also demonstrated the application of basic load types such as uniformly distributed load, as well as axial/transverse bending/torsional loads. Truss and beam elements provide a good entry point for exploring finite element simulations of components. However, the theoretical basis of these elements restricts their application to the analyses of components characterized by having a geometric length that is far greater than the dimensions of the cross-section. Essentially, they are good for long, straight one-dimensional structures, but they fall short for the analysis of many forms of advanced components.

In this chapter, we'll learn how to simulate components with more robust elements within the SOLIDWORKS simulation. Specifically, in this chapter, we will examine the **shell element** and the **axisymmetric plane element**. We will demonstrate how to use these elements to simulate axisymmetric bodies. In doing so, we will highlight the use of pressure load and load in the form of centrifugal speed within the SOLIDWORKS simulation environment. For this purpose, we will cover the following main topics:

- Overview of axisymmetric body problems

- Strategies for analyzing axisymmetric bodies

- Getting started with analyzing a thin-walled pressure vessel in a SOLIDWORKS simulation

- Plane analysis of axisymmetric bodies

Technical requirements

To complete this chapter, you will need to have access to the SOLIDWORKS software with a SOLIDWORKS simulation license.

You can find the sample files of the models required for this chapter here: `https://github.com/PacktPublishing/Practical-Finite-Element-Simulations-with-SOLIDWORKS-2022/tree/main/Chapter05`

Overview of axisymmetric body problems

A crucial aspect of engineering simulation is figuring out approximations that allow us to simplify specific problems without compromising the accuracy of our analysis. Designating certain problems as axisymmetric problems is one such approximation in the context of structural analysis.

In general, an axisymmetric body problem is characterized by two fundamental features:

- First, it involves a component that may be generated by revolving a curve or a plane section around an axis of symmetry *[1]* (see the *Further reading* section).

- Second, the component under question is assumed to have radially symmetric material properties, supports and loading configurations.

There are three important problems where the notion of *axisymmetric* considerations has contributed to the simplifications of analysis:

- **Pressurized Vessels**: This includes various kinds of vessels in the form of thin-walled/thick-walled cylindrical, conical, toroidal, and spherical pressure vessels.

- **Circular Plates**: This includes thin or thick circular plates that are loaded and supported axisymmetrically.

- **General Three-Dimensional Solids of Revolution**: This encompasses structures with radial symmetry that cannot be described as simple circular plates or pressure vessels.

Put together, these three classes of axisymmetric components play an integral part in various applications. For instance, pressure vessels are extensively used to contain and transport fluids in chemical/food/beverage processing plants and the oil and gas industries. Circular plates are deployed for use as pressure sensors, auto brake disks, nozzle covers, the end covers of pressurized vessels, the bases of storage silos, and more. Representative examples of the general 3D solid of revolution include bolt and nut assembly, aircraft fuselage, engine valve stems, turbine disks, flywheels, hyperbolic cooling towers, wheel-tire assembly, grinding wheels for completing machinery parts, and more. The following diagram shows some such examples:

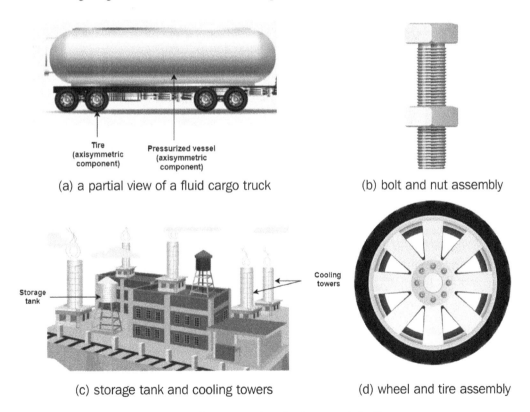

(a) a partial view of a fluid cargo truck

(b) bolt and nut assembly

(c) storage tank and cooling towers

(d) wheel and tire assembly

Figure 5.1 – Some examples of axisymmetric bodies

With a bit of background provided about axisymmetric components, let's discuss how we can analyze them.

Strategies for analyzing axisymmetric bodies

This section describes some modeling/analysis strategies for dealing with axisymmetric bodies. As you would have noticed from the previous chapters, before we can adopt a strategy for the modeling and simulation tasks, some structural details must be in place.

Structural details

The technical details that are needed for analyzing axisymmetric problems are no different from those of the previous chapters. These details have been reiterated here for completeness:

- The dimensions of the axisymmetric component should be provided.
- The material properties (the assumption of there being isotropic material properties has been adopted in this chapter as well) should be known.
- External loads must be applied to the axisymmetric component.
- Supports must be provided to the component to ensure its stability.

With the structural details sorted out, we need to make decisions about the modeling strategy to adopt. We will look at this next.

Modeling strategies

Analyzing axisymmetric bodies depends on a shared feature. This feature indicates that when these bodies are evaluated within the framework of a cylindrical coordinate system (characterized by r, θ, or z-axis), the stresses and strains that develop in these bodies during loading are independent of the circumferential coordinate θ. Bearing this in mind, any of the following modeling strategies can be adopted, depending on the type of axisymmetric body under consideration:

- **Modeling based on the full revolved body (for example, Figures 5.2a and d):** This is often very costly in terms of computational resources. For the simulation phase, the full 3D model can be analyzed by using either shell or solid elements or a combination of both. The choice of shell elements or solid elements will depend on the type of axisymmetric body under consideration.

- **Modeling based on a reduced partial, symmetric partial model (for example, Figures 5.2b, c, and e)**: By taking advantage of the axisymmetric property, this approach lowers the duration of the simulation operation. But importantly, it minimizes the number of elements required to get relatively satisfactory results. As in the first approach, the partial 3D model can be prepared for processing with either shell or solid elements or a combination of both for complicated geometries.

- **Modeling based on the use of a planar surface cross-section (for example, Figure 5.2f)**: This approach also reduces the computational resources needed to get satisfactory results. However, one of its significant differences compared to the preceding approaches lies in the use of a **two-dimensional** (**2D**) axisymmetric plane element:

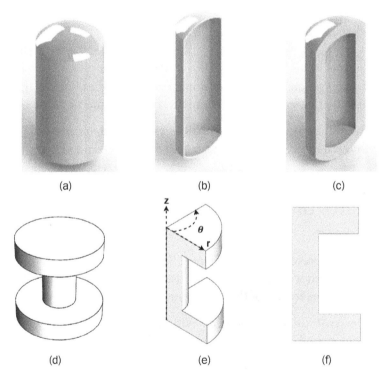

Figure 5.2 – Modeling considerations when analyzing axisymmetric bodies: (a) Full model of a pressure vessel; (b) Sectioned model of a thin-walled pressure vessel; (c) Sectioned model of a thick-walled pressure vessel; (d) Full model of a general revolved body; (e) Quarter model of a general revolved body; (f) Plane model of a general revolved body

Meanwhile, if you wish to analyze pressure vessels (and by extension, circular plate-like axisymmetric components), it is possible to break these analysis strategies into two strands:

- An analysis that employs *thin* shell elements during the simulation (for thin-walled components)
- An analysis that employs *thick* shell elements during the simulation (for thick-walled components where the effect of shear deformation is significant and needs to be considered)

From this, the following question arises: What is the key criterion that guides the decision to use thin or thick shell elements? Principally, the choice of thin or thick shell elements has to be based on the value of the ratio of the inside diameter (D_i) to the thickness (t); that is, (D_i/t). In practice, you should use the following:

- *Thin* shell elements if D_i/t > = 20
- *Thick* shell elements if D_i/t < 20

However, for analyzing general three-dimensional revolved bodies, such as the one shown in *Figure 5.2d*, you either have to use **solid elements** or axisymmetric plane elements.

We shall deal with solid elements in *Chapter 6, Analyses of Components with Solid Elements*, but for now, let's examine the key features of the shell and axisymmetric plane elements.

Characteristics of shell and axisymmetric plane elements

Two case studies will be presented in this chapter. The first will involve the use of shell elements, while the second will demonstrate the use of axisymmetric plane elements. A brief discussion about the features of these elements will be provided in this section.

The shell element is technically a 2D element. Within the SOLIDWORKS simulation library, there are two formulations of this element, as shown in the following diagram:

- The first-order triangular shell element (also called a *linear* triangular element)

- The second-order triangular shell element with six nodes (also referred to as a *parabolic* triangular element):

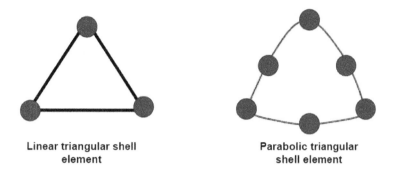

Linear triangular shell element   **Parabolic triangular shell element**

Figure 5.3 – Available shell elements in the SOLIDWORKS simulation library

Each node of the shell element shown in the preceding diagram is endowed with six degrees of freedom. These degrees of freedom are three translational displacements along the X, Y, and Z axes, and three rotational displacements about these three axes. This feature gives the shell elements the ability to resist membrane and bending loads. In practice, you will need fewer parabolic elements to accurately represent curved surfaces (compared to using linear elements).

As for the axisymmetric plane element within the SOLIDWORKS simulation library, it is similar in shape to the shell element (that is, triangular). However, it only has two degrees of freedom per node. These are principally *two translational displacements* in the plane of the element.

> **Note**
>
> There is a sound mathematical basis for these elements that can be found in many advanced textbooks on finite element analysis. There are also formulations of other non-triangular shapes in other software such as ANSYS and ABAQUS. To explore these ideas further, you may consult texts such as those by *Zienkiewicz, et al. [2], Bhatti [1],* and *Desai [3],* among others.

Now that we have highlighted the features of the elements, we will explore two case studies that will help us uncover the usage of the aforementioned elements within the SOLIDWORKS simulation environment. Along the way, some important ideas that are specific to the analysis of axisymmetric bodies will be highlighted. The next section addresses the first case study.

Getting started with analyzing thin-walled pressure vessels in a SOLIDWORKS simulation

This first case study demonstrates one of the modeling/simulation approaches for simulating thin-walled pressure vessels. As part of this demonstration, this section illustrates creating a thin surface from a thin-walled body. It also highlights how to apply pressure loads and symmetric boundary conditions. In addition, mesh refinements and methods will be provided so that you can compare the results of multiple simulation studies.

Let's get started.

Problem statement – case study 1

A thin cylindrical vessel with hemispherical heads, as depicted in the following diagram, is to be used as a gas storage tank. It is expected to withstand an internal pressure of 7 MPa. The tank is made of steel with Young's modulus of 200 GPa and a Poisson's ratio of 0.3. The internal diameter, length, and thickness of the cylinder are 75 mm, 250 mm, and 2.5 mm, respectively.

We are interested in simulating the loading of the vessel to obtain the magnitude of the principal stresses in the cylindrical section of the vessel:

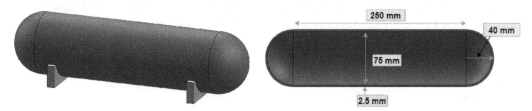

Figure 5.4 – A thin-walled pressure vessel with an internal pressure of 7 MPa

Analytical expressions exist for the principal stresses that are developed in a thin-walled cylindrical vessel with internal pressure. Nevertheless, the problem is inspired by example 9.1 in *Chapter 9, Simulation of Components under Thermo-Mechanical and Cyclic Loads*, in the book by *Hearn [4]*. In the solution, the authors obtained the values of *105 MPa* and *52.4 MPa* for two of the three principal stresses, because for thin-walled vessels, the third principal stress in the radial direction is often assumed to be negligibly small, and is hence neglected in theoretical evaluations. Note that we will be able to obtain the three principal stresses from our simulation. However, our focus is this: can we use the SOLIDWORKS simulation to verify the two principal stresses computed in *Hearn [4]*?

In an attempt to answer this question, we shall address the case study in three major sections, designated as *Part A – Creating the model of the cylinder*, *Part B – Creating the simulation study*, and *Part C – Meshing*. The first set of actions are related to creating the model of the cylinder, which is described in the next section.

Let's begin.

Part A – Creating a model of the cylinder

It is important to clarify a few points before we dig deep. First, when we simulate the problem at hand, we will ignore the effect of the support structures at the base of the vessel. Second, in practice, the vessel will have an opening (usually at the top). This will also be ignored. These two simplifications are necessary to get the result of the simulation to align with the theory of thin-walled pressure vessels *[5]* (see the *Further reading* section). In other words, we will be focusing on the resistance of the vessel's walls going forward. Consequently, we only need to create a model of the vessel.

Note

In practice, we could create a model that will contain the support structures and the openings. Then, for simplification during the simulation task, these features can be suppressed. A brief explanation of suppressing features during simulation can be found in *Chapter 6, Analyses of Components with Solid Elements*.

Furthermore, we can employ either a full model or a partial symmetric model of the vessel. Now, since we aim to demonstrate the simplification afforded by the axisymmetric condition, we shall adopt a partial symmetric model of the cylinder. Therefore, this subsection will focus on creating this partial model.

To commence, fire up SOLIDWORKS (**File → New → Part**). You are encouraged to save the file as `Cylinder` and ensure that the unit is set to the **MMGS** unit system.

Creating the revolving profile

The steps that follow aim to create the profile that will be revolved to form the geometric model of the pressure vessel. This profile is sketched on the **Top Plane** area (note that this is just a preference; you may as well use the front plane). Start by following these steps:

1. Navigate to the **Sketch** tab.
2. Click on the **Sketch** tool.

3.   Choose **Top Plane**, as shown in the following screenshot:

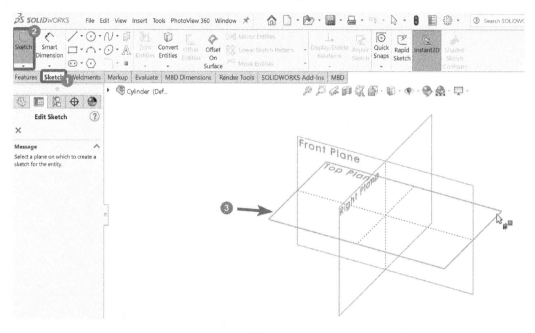

Figure 5.5 – Choosing Top Plane

4.   Use the **Line** sketching command to create the lines with the dimensions depicted in the following diagram.

Ensure that the dashed line symbolizing the axis of symmetry is positioned at the origin of the coordinate system:

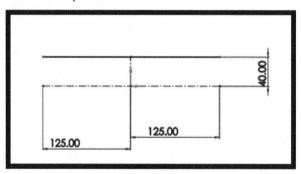

Figure 5.6 – Axis of symmetry and the generator line for the cylindrical section

5. Create two arcs on the sides of the line shown in the following diagram by using the **Centerpoint Arc** command:

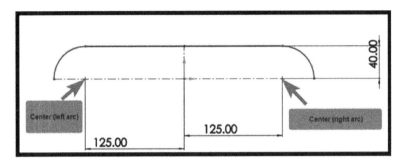

Figure 5.7 – Completing the profile of the vessel with the arc of the hemispherical head

Note that the profile we have created represents the outer edge of the vessel. Now that we've completed the profile, we can employ the revolve command.

Revolving the profile

The revolve command, like the extrude command we used in *Chapter 4, Analyses of Torsionally Loaded Components*, allows us to convert primitive line sketches into a solid body. Follow these steps to revolve the profile:

1. Navigate to the **Features** tab.

2. Click on **Revolved Boss/Base** (*Figure 5.8a*).

3. Upon activating the **Revolved Boss/Base** command, you will get the prompt window shown in *Figure 5.8b*. The prompt is asking if we want to close the profile.

   Select **No**:

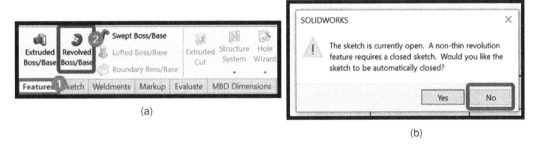

(a)

(b)

Figure 5.8 – Activating the revolve feature

At this point, we will see the **Revolve** property manager. Now, let's set some options within the **Revolve** property manager.

4. Under the **Axis of Revolution** box (labeled 1 in *Figure 5.9*), select the dotted centerline. Notice that the default value of the revolving angle (labeled 2) is **360.00deg**. If we keep this value, a full model of the vessel will be formed, as shown in the following screenshot:

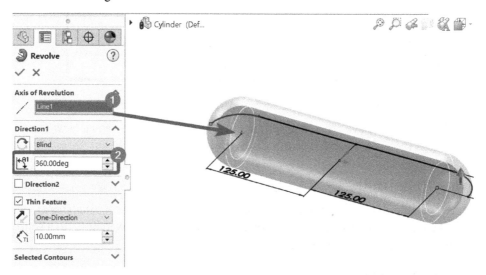

Figure 5.9 – A full model of the vessel

5. Key in the value of **90deg** in the revolving angle box (labeled 2 in *Figure 5.10a*).

6. Key in the value of **2.5mm** in the thickness box (labeled 3 in *Figure 5.10a*).

7. Click **OK** (the green ✓).

By completing these steps, you will get a quarter model of the vessel, as shown in *Figure 5.10b*:

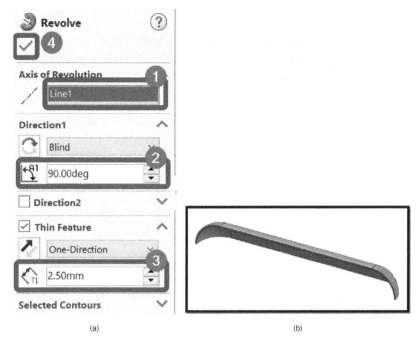

(a)   (b)

Figure 5.10 – Options for creating the quarter symmetric model of the vessel

Caution about the Revolve Feature

A tricky issue you should be mindful of when you used the revolve feature is the direction that the thickness is added. Always ensure that the side that the thickness is applied to follows your intentions. In the preceding cases, we have accepted that the thickness is to be added inside the line because the lines we created in *Figure 5.7* denote the outer wall.

It is worth pointing out that we can save even more computational resources by employing half of the quarter model we've created. If we did that, then we would have a one-eighth model, but we'll leave that as an exercise for you to try later.

Having creating the model, we are now ready to move on and convert the partial revolved body into a surface body.

Creating a surface body

You may be wondering why we need to create a surface body. The reason is that to create a shell-based mesh during the discretization process, SOLIDWORKS need a surface body. We will use the mid-surface approach of creating a surface to transform the solid revolved body into a surface body.

To this end, follow these steps to create the mid-surface body from the partially revolved body we created in the previous section:

1. Click **Insert** by going to **Main Menu** (*Figure 5.11*).

2. Choose **Offset**:

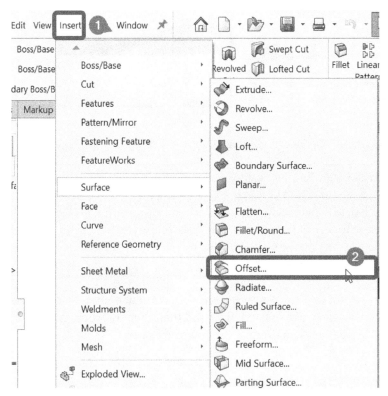

Figure 5.11 – Activating the Offset command to create a surface

The **Offset Surface** property manager will appear. Complete the necessary actions by following the following steps.

3. Click inside the **Surfaces of Faces to Offset** box (labeled 1 in *Figure 5.12*) and navigate to the graphic window to select all three inner faces.

4. Within the **Offset Distance** box (marked 2 in *Figure 5.12*), key in 1.25mm.

    If need be, click to change the direction of the offset so that the offset's surface is between the inner and outer faces.

5. Click **OK**:

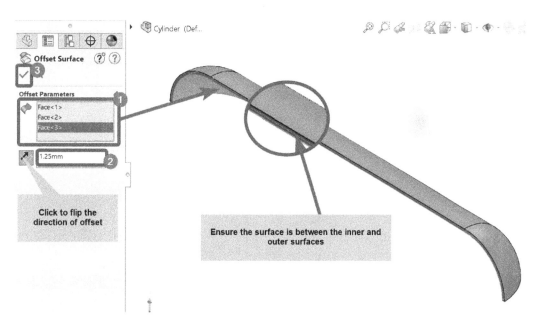

Figure 5.12 – Converting the revolved body into a mid-surface body

We now have two bodies in the form of the mid-surface body we just created and the solid revolved body from which it is formed. The folders referring to these two bodies are highlighted in *Figure 5.13a*. We will delete the solid body (in a way, it is just being suppressed).

To delete the solid body, navigate to **FeatureManager** and follow these steps:

1. Right-click on **Solid Bodies(1)**, as shown in *Figure 5.13a*.

2. Select **Delete/Keep Body**.

3.   From the **Delete/Keep Body** property manager that appears, keep the options shown in *Figure 5.13b.*:

Figure 5.13 – Deleting the solid body

The preceding steps will produce the thin surface body shown here:

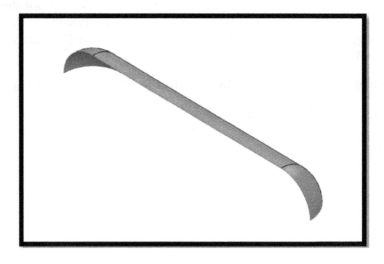

Figure 5.14 – The generated mid-surface body

Before we move on, let's add an axis that will pass through the center of the surface body. This will coincide with the axis of symmetry. Follow these steps:

1. Navigate to **FeatureManager** and select the **Front Plane** and **Top Plane** symbols, as shown in *Figure 5.15*.

2. Move to **Command Manager** and click on **Reference Geometry** (labeled 2).

3. Select **Axis** from the pull-down menu that appears:

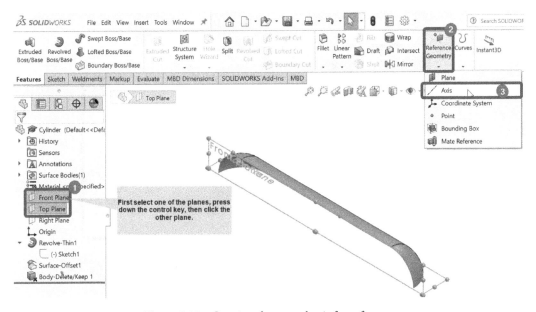

Figure 5.15 – Creating the central axis for reference

After completing *step 6*, the **Axis** property manager will appear. Keep the options within the **Axis** property manager as shown in *Figure 5.16a*. In response to the preceding actions, an axis will be created, as reflected in *Figure 5.16b*:

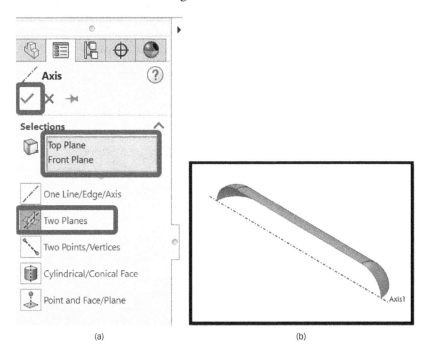

(a)   (b)

Figure 5.16 – (a) The axis property manager; (b) the created axis

At this point, we have created a mid-surface quarter model of the vessel from a partially revolved solid body. In the next subsection, we will launch the simulation study.

Part B – Creating the simulation study

This section comprises subsections that cover the usual preprocessing tasks, including activating the simulation study, assigning a material, applying a fixture, and specifying the load.

As usual, we will begin by activating the simulation environment.

Activating the simulation tab and creating a new study

To activate the simulation commands, follow these steps:

1. Click on **SOLIDWORKS Add-Ins**.
2. Click on **SOLIDWORKS Simulation** to activate the **Simulation** tab.

3. With the **Simulation** tab active, create a new study by clicking on **New Study**.

> **Note**
>
> The illustrative images for *steps 1 – 3* are similar to what we did, for instance, in *Chapter 4, Analyses of Torsionally Loaded Components,* in the *Part B – Creating the simulation study* section. If need be, you may refer to that section for a guide.

Within the **Study** property manager that appears (after completing *step 3*), follow these steps.

4. Input a study name within the **Name** box; for example, Hemi-cylinder analysis.

5. Keep the **Static** analysis option (selected by default).

6. Click **OK**.

The result of completing the preceding steps is shown in the following screenshot:

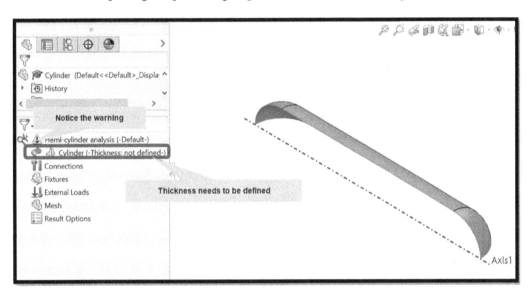

Figure 5.17 – Simulation property manager

As we can see, a warning is attached to the study name. As explained in *Chapter 4, Analyses of Torsionally Loaded Components* (in the *Converting the extruded solid bodies into beams* subsection), you can analyze a warning by right-clicking on the study name and choosing **What's Wrong**. If you do that, it will be disclosed that thickness is not defined for the shell body.

Therefore, our next action is to define the thickness of the surface.

Assigning thickness

Follow these steps to assign the thickness that will be used for the shell mesh:

1. Within the simulation study tree, right-click the surface body (named **Cylinder (-Thickness: not defined))** and select **Edit Definition**, as shown in the following screenshot:

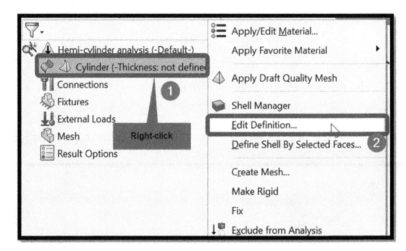

Figure 5.18 – Assigning thickness for the shell mesh

Within the **Shell Definition** property manager that appears, perform the following actions.

2. For the **Type** option, select **Thin**, as shown in *Figure 5.19.*

3. Within the thickness box (marked as 2), key in 2.5mm.

4. Activate the **Full preview** option.

5. Click **OK**:

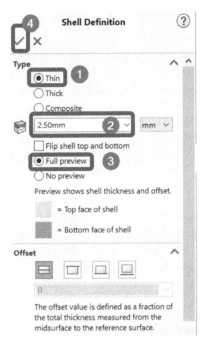

Figure 5.19 – Shell definition property manager

With that, we have defined the thickness of the surface. But notice that in addition to specifying the thickness, this step has also allowed us to select a thin shell assumption (rather than a thick shell assumption). Furthermore, once we complete the actions for defining the thickness, the warning sign attached to the study name (highlighted in *Figure 5.17*) will disappear.

Now, let's apply a material to the component.

Applying a material

By now, you should already be familiar with how to assign a specific material (due to what we explored in the previous three chapters). Nonetheless, the vessel is made of steel, according to the problem statement. To apply the desired material, follow these steps:

1. Right-click on the surface body and choose **Apply/Edit Material**.
2. Expand the `Steel` folder.

3.   Click on **AISI Type 316L stainless steel**.

4.   Click **Apply** and **Close**.

Note that the class of steel we have selected here (**AISI Type 316L stainless steel**) has Young's modulus which matches the one specified in the problem statement.

With the material assigned, our next goal is to apply the external load (internal pressure) and the specification of the fixture. Let's start by applying a fixture.

Applying a fixture

Given that we are using a partial symmetric model instead of the full model of the vessel, we need to apply symmetric boundary conditions. In this book, this is the first time that we are using this kind of boundary condition. Oftentimes, applying symmetric boundary conditions should be done with care, and there can be more than one way of doing this. For this problem, we need to apply the boundary conditions to the two major edges of the quarter model.

Follow these steps to apply symmetric boundary conditions on the edge of the quarter model that aligns with the **Front Plane** area.

Under the simulation study tree, do the following:

1.   Right-click on **Fixtures**.

2.   Pick **Advanced Fixtures** from the context menu that appears:

Figure 5.20 – Choosing Advanced Fixtures

3. Within the **Fixture** property manager that appears, select **Reference Geometry**.

4. Click inside the **Faces, Edges, and Vertices** box (labeled 1 in *Figure 5.21*). Then, navigate to the graphics window and pick the three edges that align with the **Front Plane** area, as shown in *Figure 5.21*.

5. Click inside the geometry reference box (labeled 3) and select the **Front Plane** area:

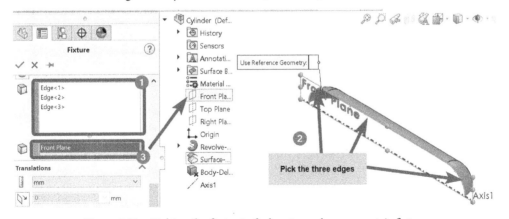

Figure 5.21 – Picking the first set of edges to apply a symmetric fixture

6. Still within the **Fixture** property manager, under the **Translations** and **Rotation** degrees of freedom, keep the options shown in *Figure 5.22.*

7. Click **OK:**

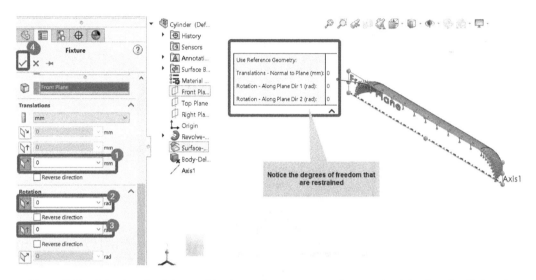

Figure 5.22 – Selecting the right combinations of movements to be restrained for the edge facing the front plane

Steps 1 – 7 covered applying the symmetric boundary conditions to the edges facing the **Front Plane** area.

> **Note about the Boundary Condition**
>
> What does the symmetric boundary condition we completed in the preceding steps imply? Well, as shown in the preceding screenshot, we have prevented the translational displacement of the edges in the direction of the **Front Plane** area. What we have done here is prevent translation of the edges along the Z-axis. Additionally, we have prevented the rotational displacements of the edges concerning **Dir 1** (which is along the X-axis) and **Dir 2** (which is along the Y-axis).

Now, we have to repeat a similar series of steps for the other symmetric edge of the model. These next set of edges align with the **Top Plane** area.

To apply the symmetric constraint to the other symmetric edge, move to the simulation study tree and follow these steps:

1. Right-click on **Fixtures**.

2. Pick **Advanced Fixtures** from the context menu that appears.

3. Within the **Fixture** property manager that appears, select **Reference Geometry**.

4. Click inside the **Faces, Edges, and Vertices** box (labeled 1 in *Figure 5.23*). Then, navigate to the graphics window and pick the three edges that align with the **Top Plane** area, as shown in *Figure 5.23*.

5. Click inside the geometry reference box (labeled 3) and select the **Top Plane** area:

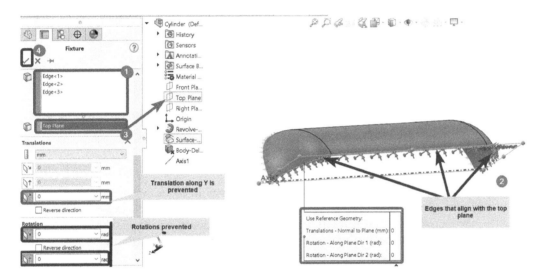

Figure 5.23 – Picking the second set of edges to apply a symmetric fixture

6. In the **Translations** and **Rotation** degrees of freedom, keep the options reflected in the preceding screenshot.

7. Click **OK**.

By completing the foregoing steps, the appearance of the model in the graphics window should look as follows:

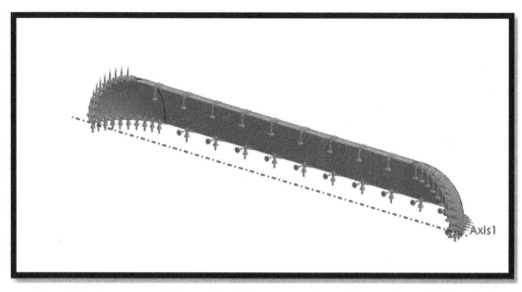

Figure 5.24 – Appearance of the model with the symmetric conditions on the two major edges

> **Note**
>
> When applying the boundary conditions to the edges of the model, we used the front and top planes as reference planes. Note that it is also possible to use the central axis we created earlier as a guide. While doing so will yield the same result, you should use reference planes to create an intuitive feel for the directions of the degrees of freedom we've constrained.

Now that we've applied the necessary symmetric boundary conditions, let's shift our attention to applying internal pressure.

Applying internal pressure

To apply the internal pressure on the inner face of the surface, follow these steps:

1. Right-click on **External Loads** and select **Pressure**:

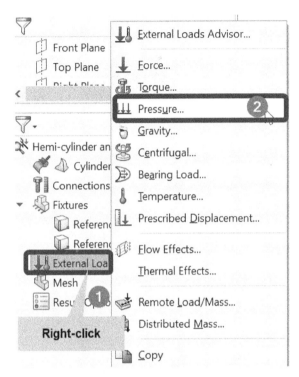

Figure 5.25 – Applying internal pressure

Within the **Pressure** property manager that appears, perform the following actions.

2. Under **Type**, keep the default option as **Normal to selected face** (labeled 1 in *Figure 5.26*).

3. Click inside the **Faces for Pressure** box (labeled 3). Then, navigate to the graphics window to pick the inner faces of the surface:

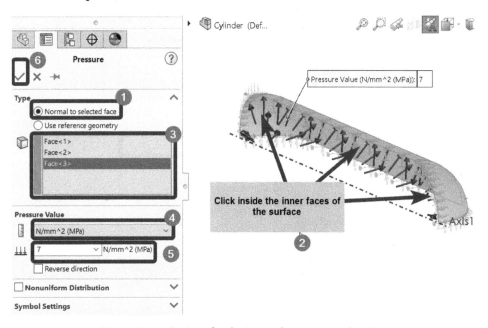

Figure 5.26 – Options for the internal pressure application

4. Under **Pressure Value**, change the unit to **N/mm^2 (MPa)** (see the box labeled 4 in *Figure 5.26*).

5. Under the **Pressure magnitude**, input **7** (ensure that the unit system is in **N/mm^2 (MPa)**, as specified in *Step 4*).

6. Click **OK**.

At this point, we have assigned thickness to the surface body, defined the material property, applied symmetric fixtures to the two major edges of the surface, and applied the internal pressure. It is now time to create the mesh.

Let's get to it.

Part C – Meshing

This section focuses on the discretization operation. Unlike in the previous three chapters, a standalone section has been dedicated to the meshing task for this problem. The reason for this hinges on the difference in the requirement for meshing components with shell-based elements (in contrast with the truss/beam-based elements).

You will also recall that, in the previous three chapters, we employed the concept of **Mesh and Run** without paying much attention to the settings of the mesh. Essentially, we exploited the concept of critical positions in those chapters, which subsequently lead to the creation of joints at critical points of the beam-like structures. Consequently, the use of critical positions eschews the need for intricate meshing operations.

However, for structures such as the pressure vessel being analyzed here, joints do not appear naturally. Simply put, these components exist as continuum structures. Consequently, the discretization of such components must be done to allow sufficient elements to be used in the analysis.

As a consequence, the general rule when using advanced elements (such as shell elements) is to pay special attention to the mesh's quality. For shell elements (and solid elements), we can experiment with the different qualitative levels of the mesh within SOLIDWORKS (for example, coarse, medium, or fine mesh). With a set of coarse meshes, we are discretizing the structure with fewer elements and thereby using fewer computational resources. In contrast, a collection of fine meshes involves a smaller number of elements. Although simulations with a finely meshed structure may take a relatively long time to run (especially for complex bodies), they produce more accurate results (assuming other aspects of the simulation workflow have been set up correctly).

In the next subsection, we will start by examining the simulation results based on the choice of a coarse mesh. Then, we will leverage a relatively fine mesh for the same problem.

We will also look at two features of SOLIDWORKS that allow a study to be duplicated and the results to be compared between different simulation studies.

Creating a coarse mesh

Begin the meshing procedure by following these steps:

1. Right-click on **Mesh** within the simulation study tree.

2. Select **Create Mesh**:

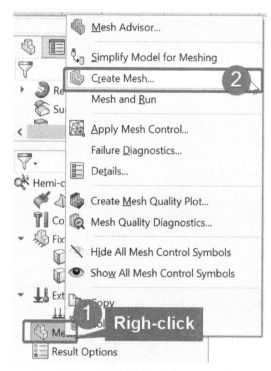

Figure 5.27 – Initiating the meshing operation

Once the **Mesh** property manager appears, you need to carry out the next steps.

3. Under **Mesh Density**, drag the **Mesh Factor** control bar (labeled 1 in *Figure 5.28a*) to the left.

4. Under **Mesh Parameters**, change the choice to **Curvature-based mesh**.

5. Click **OK**.

Figure 5.28b shows the discretized surface body using the coarse mesh settings:

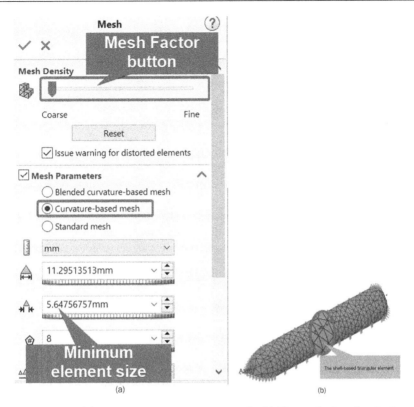

Figure 5.28 – (a) Options to create a coarse mesh; (b) The meshed surface body

Before we move on to the next action, notice that the size of the minimum triangular element (roughly 5.65 mm) highlighted in *Figure 5.28b* is bigger than the thickness (2.5 mm) of the pressure vessel being analyzed. In practice, the minimum element size should be a fraction of the thickness (or a fraction of the smallest non-aesthetic dimension) within the component. Nevertheless, despite the large size of the element, selecting **curvature-based mesh** (in *step 4*) and adopting the thin-walled option will facilitate relatively good results. Note that the curvature-based mesh employs the parabolic triangular mesh we discussed in the *Characteristics of shell and axisymmetric plane elements* subsection.

We are now set to run the analysis to obtain some preliminary results.

Running the analysis with the coarse mesh

Follow these steps to complete the running operation:

1. Right-click on the study name and select **Run** (*Figure 5.29a*).

2. You may get a suggestion about allowing the use of the **Large Displacement** option; select **No** (*Figure 5.29b*):

> **About the Large Displacement Option**
>
> Note that we have chosen to disregard the large displacement option in this analysis because we are running a linear analysis with a component that has linearly elastic material properties. The large displacement option is ideal for simulation tasks that involve nonlinear analysis or involve components with hyperelastic material properties such as soft polymers and elastomers et cetera.

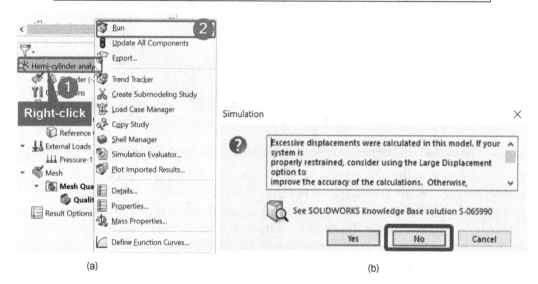

(a)   (b)

Figure 5.29 – (a) Running the analysis; (b) Disregard suggesting to use the Large Displacement option

By running the analysis, the default solution results will be produced, as shown in the following screenshot:

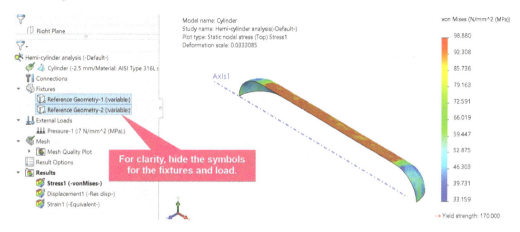

Figure 5.30 – Default results from running the analysis

As we can see, the default results include **Stress 1(-vonMises-)**, **Displacement1(-Res disp-)**, and **Strain1(-Equivalent-)**. While these values are useful for general engineering assessments, our focus for this case study is the principal stresses that were developed in the cylindrical section of the vessel. Primarily, we are interested in the first and second principal stresses so that we can compare them with the analytical results. Going forward, we shall delete the displacement and strain results and then retrieve the principal stresses.

Obtaining the first principal stress

Let's start by obtaining the first principal stress by following these steps:

1. Right-click on the Results folder and select **Define Stress plot**.

    Now, we will work with the options within the **Stress plot** property manager (summarized in *Figure 5.31a*) that appears.

2. On the **Definition** tab, navigate to **Display**.

3. Select **P1: 1st Principal Stress**.

4. Within the unit box (labeled 2 in *Figure 5.31a*), change the unit to **N/mm^2 (MPa)**.

5. For the **Shell face** option (labeled 3), select **Top** (this will display the result for the outer face of the surface).

6. Under **Advanced Options**, tick **Show plot only for selected entities** (this is labeled 4 in *Figure 5.31a*).

7. Click to activate the **Select faces for the plot** box (labeled 5 in *Figure 5.31a*). Then, navigate to the graphics window to select the outer face of the surface.

8. Now, go to the **Chart Options** tab. Then, under **Position/Format**, change the number format to **floating**, as depicted in *Figure 5.31b*.

9. Click **OK**:

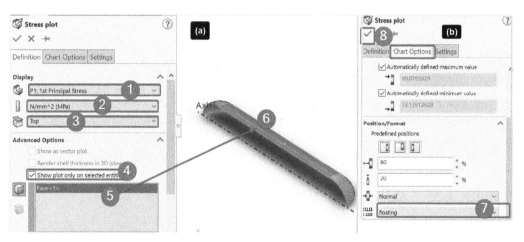

Figure 5.31 – Specifying the options for the stress plot regarding the first principal stress

With *steps 1 – 9* completed, the graphics window will be updated with the plot of the first principal stress, as shown in the following screenshot. Note that the maximum value of this principal stress happens to be roughly *114.202 MPa*. This maximum value is along the circumferential direction:

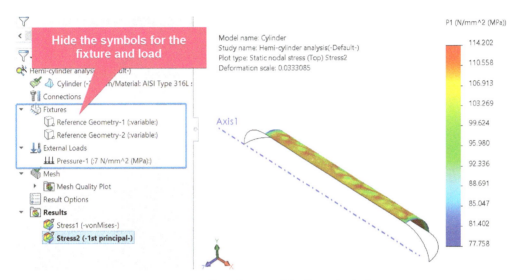

Figure 5.32 – Distribution of the first principal stress

Now, for us to compare the values of the first principal stress produced by the simulation with the analytical theoretical formula later, we must retrieve the average value of the principal stress that was developed within the cylindrical section. This should be a few points away from the edges that connect the cylinder to the hemispherical cap and away from the supported edges. To get the points that match this description, we will employ the **Probe** tool, as outlined next.

Using a probe with the first principal stress

Follow these steps to retrieve the average value of the first principal stress at specific points on the cylindrical section:

1. Right-click on **Stress2 (-1st principal-)**.

2. Select **Probe** from the resulting menu:

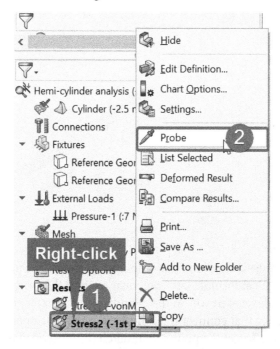

Figure 5.33 – Selecting Probe for the first principal stress

3. Keep the options in the **Probe Result** property manager that are shown in the following screenshot.

4. Navigate to the graphics window and click on four random points around the center of the top face of the surface:

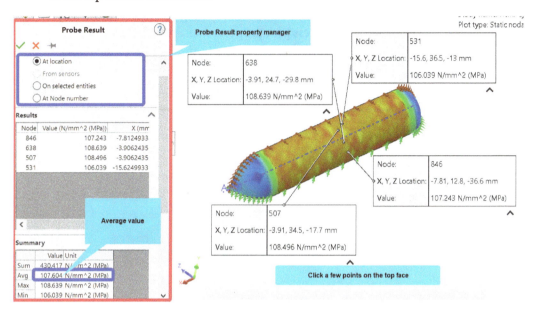

Figure 5.34 – Options for the Probe Result property manager and the average value of the first principal stress

A table summarizing the stress values will be displayed in the lower-left corner of the **Probe Result** window, as shown in the preceding screenshot. From this summary table, the average value of the first principal stress is computed to be *107.604 MPa*. Recall that we obtained the maximum value of this principal stress as *114.202 MPa* in *Figure 5.32*. We will come back to this value later. But before that, we need to obtain the second principal stress.

Obtaining the second principal stress

To obtain the second principal stress, we've got to repeat *steps 1 – 9* of the *Obtaining the first principal stress* subsection. For brevity's sake, the following screenshot illustrates the choices that need to be made via the **Stress plot** property manager:

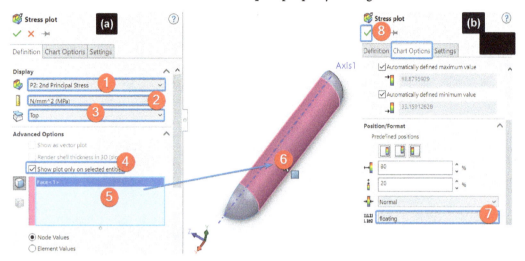

Figure 5.35 – Specifying the options for the stress plot regarding the first principal stress

Furthermore, we also need to repeat *steps 1 – 4* specified in the *Using a probe with the first principal stress* subsection. By contextualizing the aforementioned steps for the second principal stress, you should obtain the following output:

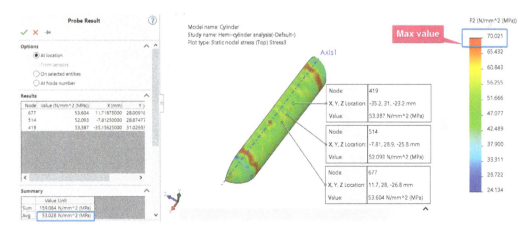

Figure 5.36 – The probed average value of the second principal stress

As we can see, the average value of the second principal stress for the cylindrical section is depicted as *53.147 MPa*. However, the maximum value of the second principal stress is determined to be *70.021 MPa*. The second principal stress corresponds with the axial/longitudinal stress. As a result, note that the maximum value of the second principal stress is located near the joint between the hemispherical end and the cylindrical sections.

We can now summarize the results as follows:

- The average value of the first principal stress: *107.604 MPa*

- The average value of the second principal stress: *53.147 MPa*

- The maximum value of the first principal stress: *114.202 MPa*

- The maximum value of the second principal stress: *70.021 MPa*

According to the theoretical analytical formula for the thin-walled cylindrical shell, the two major principal stresses are known to be the circumferential/hoop stress (maximum of the two) and the axial/ longitudinal stress (about half of the hoop stress).

Hearn [4] determined the values of the hoop and axial stresses for the problem as *105 MPa* and *52.5 MPa*, respectively. Compared to what we have just obtained from the simulation, we can conclude that the simulation result is closer to the analytical values obtained by *Hearn [4]*.

Maximum Stress Values

Note that we used the simulation to obtain the preceding stresses and we have compared some of the values with the theoretical predictions. As you may recall, at the beginning of the case study, we stated that the effect of the support structures and opening of the vessels are neglected. If we consider these components, then the location of the maximum stress may shift to the vicinity of these discontinuities. Consequently, the principal stresses from the simulation could have differed from the theoretical calculations. One of the beauties of using finite element tools such as SOLIDWORKS simulation is being able to deal with such complicating effects with ease. Simulation tools allow us to go beyond what we can do with hand calculations!

So far, we have produced results by using a coarse mesh. Now, let's improve the mesh's quality and see what happens to the maximum values of the two principal stresses. We will do this briefly in the next subsection.

You are encouraged to save the analysis file, but do not close the file yet.

Employing a finer discretization

Now, to employ finer mesh elements to rerun the analysis, we will duplicate the study we have completed using the coarse mesh. Doing so will help us build on the settings we have implemented previously, and thus save us from having to start from scratch.

To duplicate the study, follow these steps:

1. Navigate to the base of the graphics window and right-click on the study tab (named **Hemi-cylinder Analysis**).

2. Select **Copy Study**.

3. Within the **Copy Study** property manager that appears, provide a name for the new study such as Fine Mesh Analysis, as indicated in *Figure 5.37b*.

4. Click **OK**:

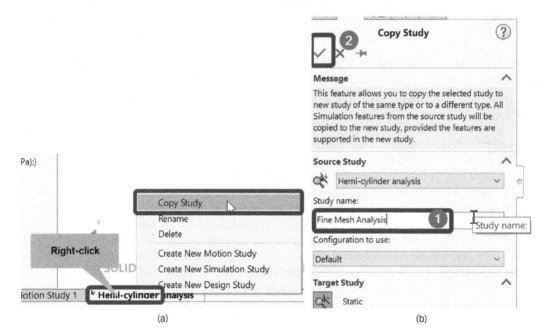

Figure 5-37 – (a) Duplicating the completed study; (b) Copy Study options

After completing *steps 1 – 4*, a new study tab will be launched beside the old study tab. Ensure you remain in this new study environment.

To refine the mesh, use the **Create Mesh** option (for a guide on this, refer to *Figure 5.27*) to generate a new mesh setting. Simply drag the **Mesh Factor** button toward the right, as shown in *Figure 5.38a*. The resulting meshed body, with a finer set of elements, should be similar to what we have in *Figure 5.38b*:

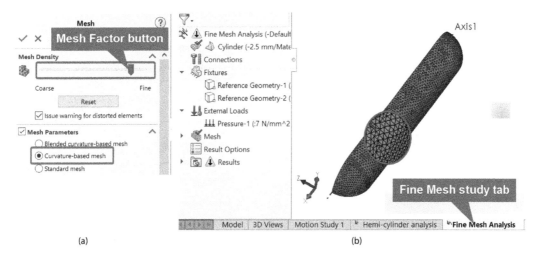

(a)   (b)

Figure 5.38 – Creating a finer mesh

> **Note**
>
> Any time you try to create a new mesh inside a duplicated study environment, you will likely get a warning stating that **Remeshing will delete the results for the study**. Since the results of the simulation with the fine mesh need to be updated, you should simply accept the warning by clicking **OK**.

Having refined the mesh, you can now run the analysis again (for a guide, refer to the *Running the analysis with a coarse mesh* subsection). Once we've done this, we will have a new set of results. Therefore, we are now in a state to compare the results.

Comparing the results

One of the useful features of the SOLIDWORKS simulation is the ease with which the results of different studies can be compared. To exploit this feature to compare the results of the analyses conducted with the finer mesh and the one conducted with the coarse mesh, follow these steps:

1. Right-click on the **Results** folder in the current study tab.

2. Select **Compare Results** (*Figure 5-39a*).

3. Within the **Compare Results** property window that appears, under **Options**, tick the **All studies in this configuration** box (this action reveals all the available studies in the configuration).

4. To compare the values of the two principal stresses from the two studies, make the selection depicted for the box labeled 2 in *Figure 5.39b*:

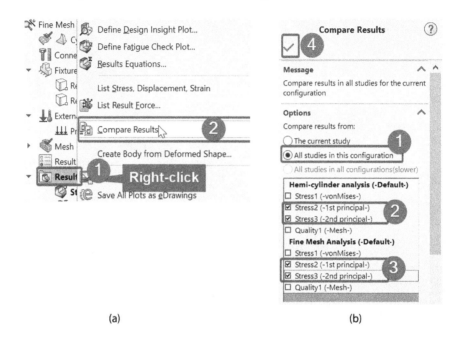

(a)   (b)

Figure 5.39 – Initiating the process of comparing the results

Upon completing the preceding actions, the graphics window will be updated and four windows corresponding to the four principal stresses will be displayed, as shown in the following screenshot:

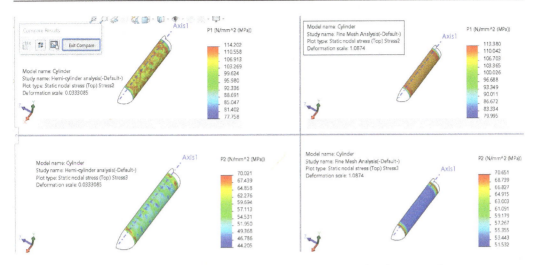

Figure 5.40 – Four windows showing the selected results for the two studies

As you can see, the maximum value of the first principal stress has reduced to *113.382 MPa* (which is roughly 5.4% closer to the theoretical prediction), while the minimum value of the second principal stress has increased to *51.532 MPa* (roughly about 3.4% of the value of the second principal stress). Further, you will notice that there is a smooth variation in the color of the stress distribution for the results obtained using the fine mesh, indicating much better convergence of the results.

For this simple problem, the effect of refining the mesh may not appear to be significant, but for a large assembly of components, analysts must carry out convergence analysis to determine how changes in mesh quality affect a specific result of interest (usually stress or displacements in static analysis) and to determine an acceptable element size. The reason for this is that, oftentimes, for a large assembly with multiple components, using a coarse mesh may not always give an accurate result, while using an ultra-fine mesh will incur a costly simulation duration. So, a compromise is needed to strike a good balance between mesh quality, the desired result accuracy, and simulation duration. We will explore convergence analysis in more detail in *Chapter 10, A Guide to Meshing in SOLIDWORKS*.

We hope this section has given you a taste of what effect the mesh's quality can have when using advanced elements such as shell elements. At this point, it is worth pointing out that although we have focused our attention on the principal stresses, you can pretty much retrieve other results of interest. For instance, we could have also used the value of the von Mises stress to determine if the vessel will fail or not. We could have also used the value of the resultant displacements to determine if the pressure vessel experiences excessive deformation. All these are left for you to explore.

This concludes the simulation of the first case study. This specific case study has taken us on a journey that has seen us using the shell element, applying symmetric boundary conditions, creating and refining shell meshes, and comparing results across two studies.

Now, because we will not be returning to the topic of axisymmetric body problems again in future chapters, you may want to explore one more case study, even if briefly, that will not be based on the use of shell elements.

Plane analysis of axisymmetric bodies

The approach that was adopted in the previous case study involved a thin-walled pressure vessel, for which the use of shell elements is more suitable. However, there are a variety of axisymmetric problems that involve engineering components with substantial thickness. In such cases, the use of thin shell elements may not be the best choice. This section relates to one such case.

Problem statement – case study 2

Let's start with a bit of background for the case study. In power presses, automobile engines, and a wide variety of power generation/transmission systems, mechanical flywheels are used as part of the energy storage system to smoothen out power delivery [6]. Indeed, recently, the design of the flywheel has taken a central position with the recent shift in consciousness to the demand for renewable energy. In its most basic form, a flywheel unit consists of a shaft and a solid rotating disk acting as the rotor (see *Figure 5.41*). One potential source of concern in the design of a flywheel energy storage system is the failure of the rotating disk due to high speed:

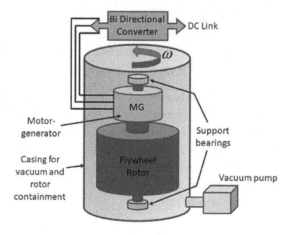

Figure 5.41 – Flywheel unit for electrical energy storage (see Amiryar and Pullen [7])

In this case study, we will assess the von Mises stress as well as the radial and tangential stresses that develop in a flywheel disk rotating with a mid-range speed of *20,000 rpm*. *Figure 5.42a* shows a typical flywheel disk with a partial view of the power transmitting shaft. A sectioned view of the disk is shown in *Figure 5.42b*:

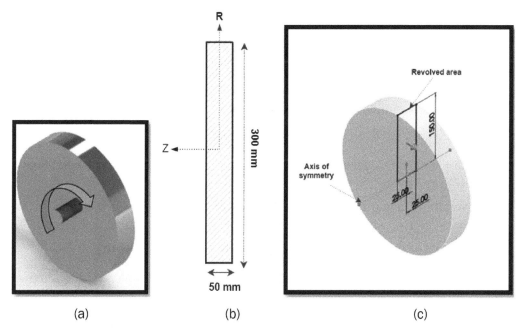

Figure 5.42 – (a) Flywheel rotor; (b) The right plane sectioned view; (c) A simplified rotating disk showing the revolved area

However, for us to compare the value of the stress obtained from our simulation with the analytical solutions for the radial and tangential stresses of a solid rotating disk, we will suppress the effect of the protruded shaft and concentrate on the disk, as shown in *Figure 5.42c*. Meanwhile, also shown in *Figure 5.42c* is the plane rectangular area that may be revolved to form the flywheel. The flywheel is made of steel with Young's modulus of 200 GPa and a Poisson's ratio of 0.3.

A brief guide to the simulation of this problem is given next.

Modeling the flywheel

In principle, there are a few ways to solve this problem using the finite element method:

- Use a simplified revolved plane section.

- Use a partial symmetric solid model, which will exploit the rotational symmetry of the problem.

- Use the full solid model of the solid disk.

The latter two approaches will require the use of solid elements, which will be covered in *Chapter 6*, *Analyses of Components with Solid Elements*. So, for now, we shall go with the approach of using the simplified revolved plane section.

A compressed set of steps to create the model are as follows:

1. Create a model of the rectangle that serves as the revolved plane on the **Right Plane** area (*Figure 5.43a*).

2. Convert the area into a planar surface by going to **Insert → Surface → Planar** (*Figure 5.43b*).

3. Add an axis to the base of the surface by going to **Reference Geometry → Axis** (*Figure 5.43c*):

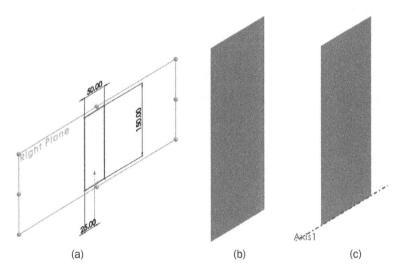

Figure 5.43 – Creating the planar surface

Later in the simulation environment, we will want to apply boundary conditions to stabilize the finite element model. For this, we need to use a vertex at the middle of the edge at the base of the surface. As-is, the created surface has only four vertices at the corners. One way to create a vertex at the middle base edge of the surface is to use the **Split** tool. Follow these steps to split the surface:

1. Navigate to the **Main Menu** pulldown, select **Insert → Features → Split**.

2. In the **Split** property manager that appears, click inside the **Trim Tools** box (labeled 1 in *Figure 5.44*). Then, move to the feature manager tree to select the **Front Plane** area.

3. Click **Cut Part**.

4. Under **Resulting Bodies**, tick inside the boxes, as shown in the box labeled 3 in *Figure 5.44*.

5. Click **OK**:

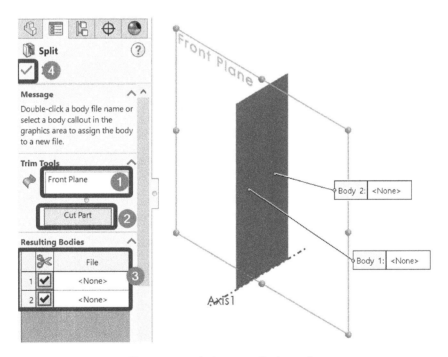

Figure 5.44 – Actions to split the surface

Once you've completed *steps 1 – 5*, a symbol representing that the body has been split should appear in the feature manager, as indicated in *Figure 5.45*. Notice the presence of the vertex in the model:

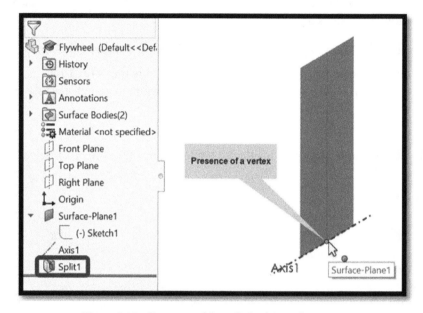

Figure 5.45 – Presence of the split bodies and a vertex

Now, our model is ready for simulation, which is what we will shift our attention to next.

Simulating the flywheel

The steps to create a new study based on 2D simplification are different from what you have seen previously. For this purpose, carefully follow these steps:

1. Using the **Simulation** tab, launch a **New Study**.
2. From the **Study** property manager that opens, provide a name for the study, such as `Flywheel Plane Analysis`.
3. Tick the **Use 2D Simplification** box, as shown in *Figure 5.46a*.
4. Click **OK**.

Now, unlike all the previous studies/analyses we have conducted so far, by clicking **OK** another window – the **2D Simplification** property manager – will open. This will allow us to pick the right type of planar simplification we wish to deploy.

5. Within the **2D Simplification** property manager, under **Study Type**, choose **Axi-symmetric** (see *Figure 5.46b*):

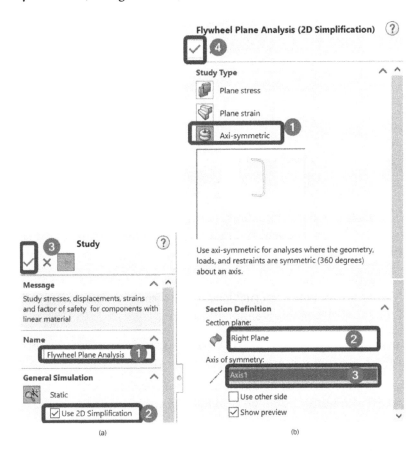

Figure 5.46 – Selecting the options for a 2D plane axisymmetric analysis

6. Under **Section Definition**, choose the **Right Plane** area from the feature manager.

7. For **Axis of Symmetry** (labeled 3 in *Figure 5.46b*), choose **Axis1**.

8. Click **OK**.

Note we have selected the **Right Plane** area in *step 6* because we created the revolved area on this plane. Upon completing *steps 1 – 8*, we will get the now-familiar simulation study tree shown in the following screenshot:

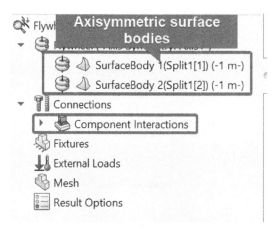

Figure 5.47 – Simulation study tree for planar analysis

Note the symbol attached to the surface body in the preceding screenshot. You will also notice the presence of the **Component Interactions**. This arises because of the splitting operation we performed earlier.

Now that we have arrived at the simulation environment, it is time to assign a material property, apply a fixture, apply the external load in the form of a rotational speed, create the mesh, and run the analysis. This is summarized here:

1. For the material, assign **AISI Type 316L stainless steel** (for a guide on this, refer to the *Applying a material* subsection for case study 1).

2. To apply the fixture, begin as we did in the *Applying a fixture* section. Using those steps, specify a **Roller/Slider** support on the edge at the base and use the options shown in *Figure 5.48a*:

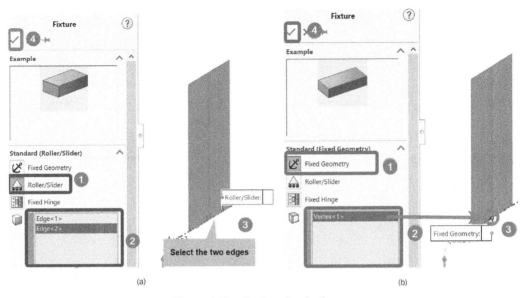

Figure 5.48 – Options for the fixtures

3. For the load application (centrifugal speed), follow the steps illustrated in
    Figure 5.49:

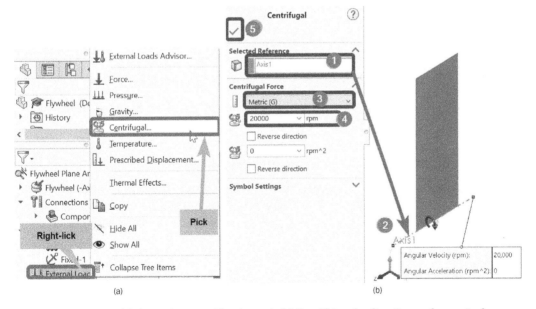

Figure 5.49 – (a) Activating centrifugal speed; (b) Specifying the direction and magnitude
of the centrifugal speed

4. To create the mesh, follow the steps outlined in the *Creating a coarse mesh* subsection. However, instead of using a coarse mesh, let the **Mesh Factor** button be in the middle. This will create a mesh of medium quality that should be satisfactory.

    At this point, the simulation study tree, together with the meshed body in the graphics area, should appear as follows:

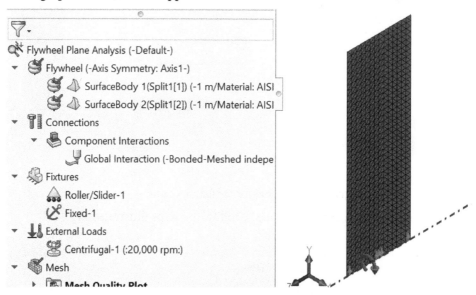

Figure 5.50 – The updated simulation study tree and the mesh's planar body

5. Run the analysis (right-click on the name of the analysis and choose **Run**).

By completing all the actions outlined in *steps 1 – 5*, we are in a position to explore the results,.

Extracting the principal stresses in the flywheel

For validation purposes, we will extract the first two principal stresses. But before we obtain these results, let's estimate the expected value of the radial and the tangential stresses that will develop in the disk, as predicted by the following theoretical formula (see *Barber [8]*):

$$\sigma_r = \frac{\rho\omega^2(3+v)}{8}\left(r_i^2 + r_o^2 - \frac{r_i^2 r_o^2}{r^2} - r^2\right)$$

(5.1)

$$\sigma_t = \frac{\rho\omega^2(3+v)}{8}\left(r_i^2 + r_o^2 + \frac{r_i^2 r_o^2}{r^2} - \frac{1+3v}{3+v}r^2\right)$$

(5.2)

Here, r is a varying radial distance, while r_o and r_i denote the external and internal diameters, respectively. Furthermore, ρ, v, and ω symbolize density, Poisson's ratio, and angular velocity, respectively. From the problem statement and the material property values from the SOLIDWORKS library, we get the following formula:

$$\rho = 8027 \frac{kg}{m^3}, r_i = r = 0, r_0 = 150 \, mm, v = 0.265; \omega = 2\pi(20000)/60$$

Using these values, we can determine the stresses as follows:

$$\sigma_r = \frac{8027 \, (3 + 0.265)}{8} \left(\pi \times \frac{2000}{3}\right)^2 (150 \times 10^{-3})^2 = 323.33 \, MPa \tag{5.3}$$

$$\sigma_t = \frac{8027 \, (3 + 0.265)}{8} \left(\pi \times \frac{2000}{3}\right)^2 (150 \times 10^{-3})^2 = 323.33 \, MPa \tag{5.4}$$

By obtaining the principal stresses (previously illustrated and discussed for case study 1), you can compare the results from the simulation with the stress values we obtained from the analytical formula. The following screenshot shows the two principal stresses from the simulation:

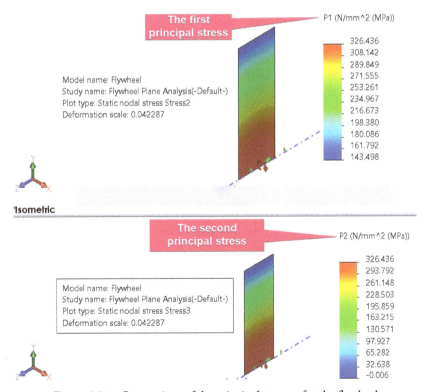

Figure 5.51 – Comparison of the principal stresses for the flywheel

As you can see, we have obtained a value of *326.436 MPa* for **P1** (the upper screenshot) which denotes the first principal stress, and the same value for **P2** (the lower screenshot in *Figure 5.51*). You will recognize that a negligibly small difference exists between the computed results we have obtained from the SOLIDWORKS simulation and the prediction from the analytical expressions. This has happened despite the fact we have used fewer axisymmetric plane elements compared to what we would have used if we had conducted the analysis using the whole three-dimensional model of the flywheel disk.

The last result we will look at is the von Mises stress, which is often used as an indicator to determine the survival of the component under the applied speed. In retrieving the value for the von Mises stress, we will also show another interesting aspect of the stress plot property when using the axisymmetric plane elements, which is obtaining a *3D stress plot* to perform an analysis with the axisymmetric plane elements.

Recall that the von Mises result is part of the default solution that is available under the **Results** folder. Consequently, what we need to do is simply edit the settings of the computed von Mises stress to obtain the 3D plot of its variation. For this purpose, follow these steps:

1. In the **Results** folder, right-click on **Stress 1 (-von-Mises-)** and choose **Edit Definition** (*Figure 5.52a*).

2. Within the **Stress** plot property manager that appears, choose the options indicated in *Figure 5.52b*. Ensure that you tick the **Show 3D plot** box (labeled 3). You can also specify any desired value in the angle box (labeled 4).

3. Click **OK**:

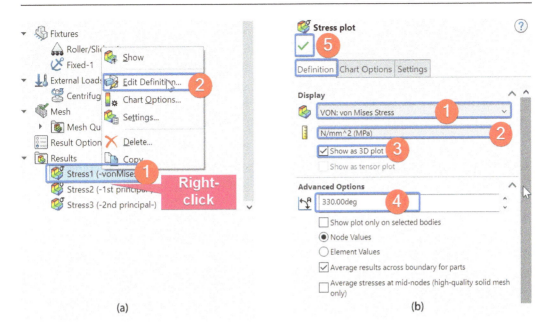

Figure 5.52 – (a) Editing the plot of the von-Mises stress; (b) Ways to obtain
a 3D plot of the von-Mises stress

Upon completing *steps 1 – 3*, the graphics window will be updated to display the
variation of the von Mises stress, as depicted in the following screenshot. Here, both the
maximum von Mises stress and the **Yield strength** property of the material of the disk are
highlighted:

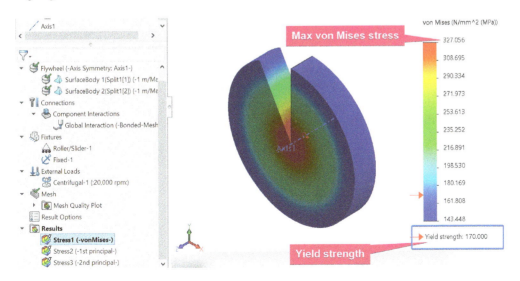

Figure 5.53 – The 3D plot of the von-Mises stress

The preceding screenshot indicates that the maximum von Mises stress is way beyond the **Yield strength** property of the disk. This shows that the disk will not survive the operating speed of 20,000 rpm that has been applied to it during operation. To improve this design, you may wish to change the material increase the thickness of the disk. Alternatively, if you are dealing with a design task for which the material specification is fixed, then an optimization task can be conducted to determine the optimal thickness that will allow the disk to operate safely.

With that, we can conclude this second case study. You will agree that we have come a long way. We started by providing some background details about three-dimensional bodies that are endowed with axisymmetric properties. We then walked through our first case study, which dealt with the static analysis of a thin-walled cylinder using shell elements. Following that, we took on the second case study and explored the use of the **plane axisymmetric element** to analyze a flywheel idealized as a rotating disk. Note that although we have deployed shell elements in this chapter solely for analyzing axisymmetric problems, their usage goes far beyond this. For instance, they can be used to analyze various forms of non-axisymmetric, thin-walled structural members.

Analysis of thick-walled cylinders

The method of using plane axisymmetric plane elements that we have exploited for the flywheel case study can also be used to conduct a planar analysis of thick-walled cylinders. But in most practical simulation problems, solid elements may be more ideal for analyzing thick-walled cylinders.

Summary

In this chapter, we explored the procedures involved in analyzing axisymmetric bodies. Various strategies for analyzing components that satisfy the axisymmetric condition were highlighted. Using two case studies, we have demonstrated how to analyze axisymmetric components with shell elements (thin-walled members) and with the axisymmetric plane element. A few of the key concepts we explored included:

- How to apply symmetric boundary conditions to the edges of partial symmetric surface body
- How to apply a pressure load and centrifugal speed load
- How to duplicate a study and how to create/refine the simulation mesh

- How to compare results between studies in the same configuration

- How to use the probe tool to obtain the average values of stress

- How to visualize a 3D plot for a study conducted with an axisymmetric plane element

You can deploy these concepts for similar design problems that you may have. In the next chapter, you will learn about how to analyze components with solid elements.

Exercises

1. A thin-walled cylindrical vessel that's 3 meters in length has an internal diameter of 1 meter and a thickness of 30 mm. The cylindrical vessel is made of AISI Type 316L stainless steel and it will contain a fluid pressure of 2.1 MPa. Create a quarter model of the cylinder and use the SOLIDWORKS simulation to determine the hoop and axial stresses in the cylinder.

2. An aircraft cabin window has been designed in the form of a thin circular plate with a radius of 600 mm and a thickness of 20 mm. The material of the window is polycarbonate with Young's modulus of 13.5 MPa and a Poisson's ratio of 0.36. Create a full model of the plate and analyze its maximum deflection and maximum principal stress using shell elements.

3. Reconduct case study 2 for a disk that has the same diameter but with a thickness of 100 mm. Use SOLIDWORKS to determine the factor of safety for the new disk under the same operating speed of 20,000 rpm and the same material property.

Further reading

- [1] *Advanced Topics in Finite Element Analysis of Structures: With Mathematica and MATLAB Computations, M. A. Bhatti, Wiley, 2006.*

- [2] *The Finite Element Method: The basis, O. C. Zienkiewicz, R. L. Taylor, R. L. Taylor, and R. L. Taylor, Butterworth-Heinemann, 2000.*

- [3] *Finite Element Method with Applications in Engineering, Y. M. Desai, Dorling Kindersley, 2011.*

- [4] *Mechanics of Materials Volume 1: An Introduction to the Mechanics of Elastic and Plastic Deformation of Solids and Structural Materials, E. J. Hearn, Elsevier Science, 1997.*

- [5] *Pressure Vessel Design, D. Annaratone, Springer Berlin Heidelberg, 2007.*

- [6] *Power system energy storage technologies, P. Breeze, Academic Press, 2018.*

- [7] *A review of flywheel energy storage system technologies and their applications, M. E. Amiryar and K. R. Pullen, Applied Sciences, Vol. 7, no. 3, p. 286, 2017.*

- [8] *Intermediate Mechanics of Materials, J. R. Barber, Springer Netherlands, 2010.*

6
Analysis of Components with Solid Elements

This chapter builds on the idea of advanced elements within the SOLIDWORKS simulation library that we commenced looking at in the preceding chapter. Primarily, the current chapter hones in on **solid elements**, versatile elements for the simulation of arbitrary three-dimensional bodies. In the course of exploring these elements with two concrete examples, the chapter features the use of mesh control, the setting up of global/local interactions between components, the significance of curvature-based meshes, and an assessment of contact stress, among other things. Overall, the following topics will be covered:

- Overview of components that deserve to be analyzed with solid elements
- Analysis of helical springs
- Analysis of spur gears

Technical requirements

You will need to have access to the SOLIDWORKS software with a SOLIDWORKS Simulation license.

The sample files used in this chapter can be found here:

```
https://github.com/PacktPublishing/Practical-Finite-Element-
Simulations-with-SOLIDWORKS-2022/tree/main/Chapter06
```

Overview of components that deserve to be analyzed with solid elements

As an analyst, you will be confronted with a wide range of simulation tasks, and the onus will be on you to decide on the most appropriate element as part of the finite element simulation workflow. Choosing the right element has an enormous impact on the accuracy of your simulations.

In the context of static analysis, the solid element remains one of the most flexible elements suitable for the analysis of various kinds of components. In the previous chapters, we mentioned some limitations of the truss/beam/shell elements when they were introduced. Although these elements allow us to use fewer elements in our simulations, they are all mathematical and mechanistic approximations of three-dimensional (3D) bodies. Therefore, in practice, a set of solid elements can be used whenever the structure you wish to analyze violates the limitations of structural elements such as beam/shell elements. For instance, anytime the dimensions of components are comparable to each other, then adopting solid elements for discretization becomes necessary (although not sufficient).

The set of engineering components that can be sufficiently analyzed with solid elements is vast. This ranges from simple to geometrically complex engineering structures and machinery. *Figure 6.1* shows some examples of components that can be analyzed with solid elements:

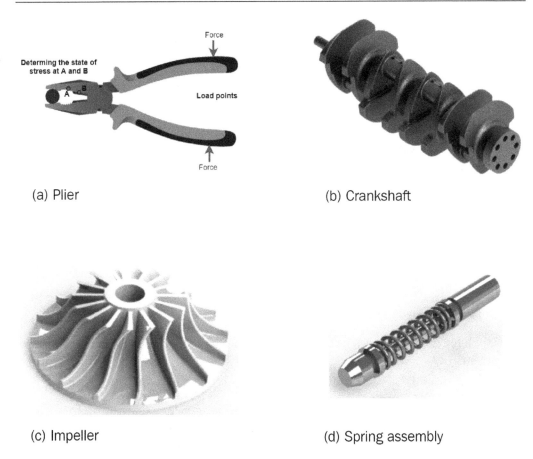

(a) Plier   (b) Crankshaft

(c) Impeller   (d) Spring assembly

Figure 6.1 – Some examples of structures that may require solid elements for analyses

Among the group of seemingly simple components that needs to be analyzed with solid elements is the following non-exhaustive list:

- Non-prismatic beams with variable geometric details
- Short and thick beams/shafts/plates/panels with structurally significant stress raisers in the form of perforations, cut-outs, fillets/chamfers, grooves, keyways, and so on
- Thick-walled tanks, pipes, and cylinders with complicated connections and support effects
- Axisymmetric bodies with non-symmetric loads/supports/material properties
- Machine elements in the form of springs, gears, and so on

Examples of structures within the category of geometrically complex engineering components and machinery are far too numerous to list. A few examples include machinist vise, T crane hook, crankshaft, pliers, impeller, heavy-duty lifting brackets, and so on.

Now that we have gotten the basic background out of the way, it is time to bring into the picture some details and strategies necessary for the analysis of general 3D bodies.

Structural details

The primary technical details needed for the analysis of 3D bodies are again similar to those we listed in the previous chapters. These include the following:

- The dimensions of the component (for an assembly, the dimensions of all members)
- Material properties (the assumption of the isotropic material property is again adopted in this chapter)
- External loads applied to the component/assembly
- Supports provided to the component/assembly

Next, we will touch upon a few tricks to simplify or modify components for analysis with solid elements.

Model simplification strategies

It is impossible to have a list of all-encompassing strategies to be adopted for the simplification of components to be analyzed with solid elements. Nevertheless, three basic strategies that have wide applications are listed here:

- *Model simplification via suppression*: This strategy involves suppressing features that perform purely aesthetic or less structural functions before launching the simulation study environment. Such features comprise fillets, chamfers, decals/embossed text, cosmetic threads, and so on. For instance, the fillets and chamfers in *Figure 6.2a* have been suppressed in *Figure 6.2b*:

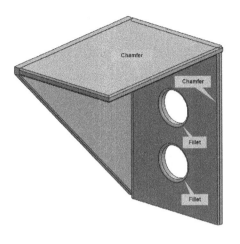
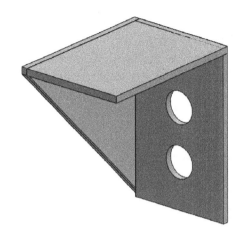

(a) Original geometric model   (b) A simplified version of the model

Figure 6.2 – An illustration of the suppression of fillets and chamfers

Note that feature suppression can be done manually or with the automated option within the simulation environment.

- *Defeaturing*: For very large components with a horde of small features and small aesthetic parts, the manual suppressing approach becomes time-consuming and almost impractical. Besides, for large assemblies, you may have to deal with small aesthetic features that are hidden within several connecting parts. In such scenarios, it is better to fall back on the automated approach that relies on the SOLIDWORKS **defeature** command. This can be accessed from the main menu (**Tools → Defeature**).

The two strategies outlined so far demand that we remove certain features to simplify the model, but we can also add some subtle features that will simplify the downstream finite element simulation tasks. One such kind of addition is discussed next.

- *Addition of split line/reference point/reference axis:* These three features (split line, reference point, and reference axis) are very useful for the creation of a partial face, an arbitrary point, or a desired axial line for the application of loads/supports on solid bodies. As with the previous approaches, these features are created before launching the simulation study. *Figure 6.3* illustrates two scenarios where a split line and a reference point are added to the model to allow the application of distributed and point loads:

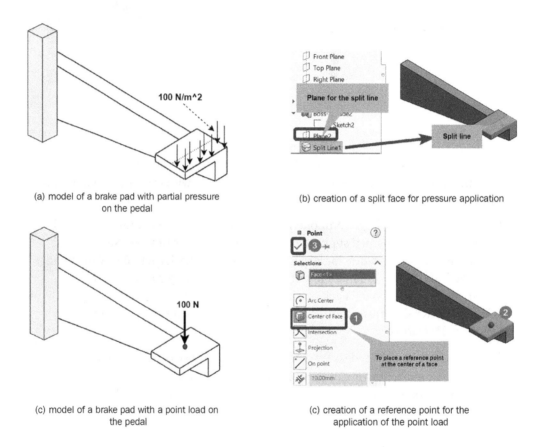

(a) model of a brake pad with partial pressure on the pedal

(b) creation of a split face for pressure application

(c) model of a brake pad with a point load on the pedal

(c) creation of a reference point for the application of the point load

Figure 6.3 – Addition of a split line/reference point for load applications

In a way, the addition of a split line, reference points, and reference axes on solid bodies mirror our use of critical positions in *Chapter 3*, *Analyses of Beams and Frames*.

Before we get to the case study, let's touch a little bit on a few attributes that solid elements possess.

Characteristics of solid elements

The first attribute is that solid elements are truly 3D sub-domains with well-defined volume data. Another is that they sit atop the family tree of continuum elements (discussed in *Chapter 1, Getting Started with Finite Element Simulation*). Furthermore, two popular categories of solid elements are widely used in finite element simulations (*Figure 6.4*): (i) the hexahedron-based solid elements – a 3D extension of the 2D quadrilateral elements that are also sometimes called brick elements; and (ii) the tetrahedron-based solid elements – a 3D extension of the 2D triangular elements. In terms of usage, in most cases, a collection of tetrahedral elements is suitable for the analysis of irregular solid bodies. However, in the absence of irregular features (such as corners, grooves, keyways, and so on), a collection of hexahedral elements will provide better accuracy and will often require less simulation runtime compared to tetrahedral elements.

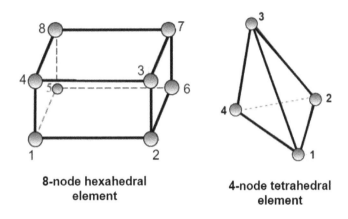

8-node hexahedral element   **4-node tetrahedral element**

Figure 6.4 – An illustration of the shapes and nodes of hexahedral and tetrahedral elements

Note that only tetrahedral solid elements are available in SOLIDWORKS.

As with the shell element discussed in *Chapter 5, Analyses of Axisymmetric Bodies*, two formulations of the tetrahedral solid element exist within the SOLIDWORKS simulation library:

- The first-order (4-node) tetrahedral solid element (also referred to as a linear solid element)

- The second-order (10-node) tetrahedral solid element (also referred to as a parabolic solid element)

Now, as a consequence of being 3D and belonging to the family of continuum elements, each nodal point of the aforementioned solid elements has only *three translational displacement degrees of freedom along the X, Y, and Z axes*. They do not have rotational degrees of freedom.

> **Note**
>
> For further explorations of the mathematical formulations of solid elements, you may consult the books by *Koutromanos [1]* (see the *Further reading* section), *Reddy [2]*, and *Hutton [3]*, among others.

Drawing on the ideas we have presented so far, we will now dig into the exploration of two case studies that will allow us to employ solid elements in our simulation. The first is on helical spring, while the second relates to spur gears. The two case studies will showcase some new concepts, for instance:

- How to create and prepare a variable pitch spring for analysis
- Recognizing the pitfalls of the standard mesh
- How to simplify a multi-body component/assembly by excluding some components from the analysis
- How to set up a "no penetration" contact and obtain contact stress

We begin with the first case study in the next section.

Analysis of helical springs

Our first case study is a simulation exercise that involves a helical compression spring. Springs are known to store energy when deflected and return the same energy upon release from the deformed state. Their use can be found in systems as complex as vibration isolation systems, shock absorbers, watches, and bomb detonation mechanisms, to simple devices such as toy cars and pens. Detailed accounts of the design of springs can be found in some of the references in the *Further reading* section, for example, *[4]*. In what follows, we will examine the simulation of a helical compression spring as part of a sub-assembly.

Problem statement – case study 1

Figure 6.5 shows different views of a spring-based weighing module. The performance of the module is intrinsically tied to the deformation of the compression helical spring (isolated in *Figure 6.5c*). As a result, understanding the performance of the spring is a necessary engineering design task. The spring is to be made from an ASTM316 stainless steel wire with a diameter of 10 mm. Other data about the spring are listed here:

- Total number of active coils = 10
- Number of inactive coils = 4
- Free length of the spring = 500 mm
- The outside diameter of the spring = 100 mm

The spring is ground at the ends. The objective is to compute the deflection and the maximum shear stress experienced by the spring when a load of 25 kg is placed on the load carrier (see *Figure 6.5b*). The load carrier weighs 500 g, resulting in a pre-load of 5 N. Further, we wish to know how well the deflection and the shear stress retrieved from our simulation compare with the theoretical expressions for springs' deflection and shear stress developed by *Wahl* [5].

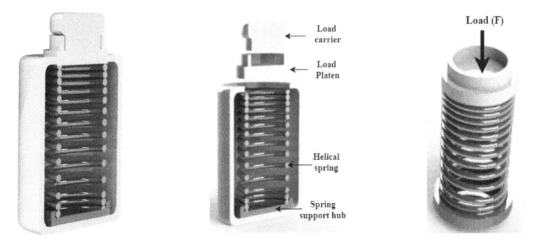

Figure 6.5 – Spring-based weighing system: (a) a sectioned view of the simplified assembly; (b) an exploded view showing the major components; (c) the spring sub-assembly

In what follows, we will walk through a systematic procedure to create and analyze the spring sub-assembly in *Figure 6.5c*. As in earlier chapters, we will address the solution to the above problem in three major sections designated as *Part A: Creating the model of the spring sub-assembly*, *Part B: Creating the simulation study*, and *Part C: Meshing and post-processing of results*.

Let's begin with the modeling phase.

Part A: Creating the model of the spring

There are three components we will use for the simulation: (a) the helical spring; (b) the spring support hub; and (c) the load platen. *Figure 6.6* shows the major geometric dimensions of these components:

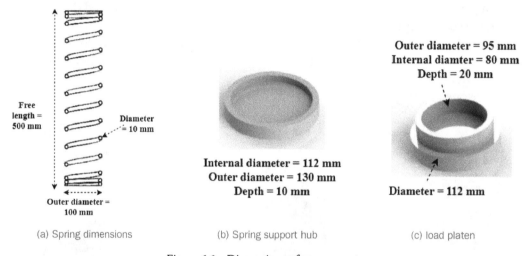

(a) Spring dimensions   (b) Spring support hub   (c) load platen

Figure 6.6 – Dimensions of components

Due to space limitations, the explanation that follows focuses squarely on the creation of the variable pitch spring (since the other components are easy to model).

To commence, start up SOLIDWORKS (**File → New → Part**). You are encouraged to save the file as Spring and ensure the unit is set to the **MMGS** unit system.

Creating the helix profile of the variable pitch spring

The backbone of the spring is the helix guide curve. The helix can be generated by specifying the following geometric details *[6, 7]*:

- A reference circle matching the diameter of the helix
- The height of the helix, which should correspond with the free end of the spring
- The pitch, which describes a distance between identical points along the axial direction of the helix
- The revolution, which describes the number of complete turns of the helix

So, let's begin by sketching the reference circle of diameter 100 mm on **Top Plane** by taking the following steps:

1. Navigate to the **Sketch** tab.
2. Click on the **Sketch** tool.
3. Choose **Top Plane**.
4. Use the **Circle** sketching command to create the circle as depicted in *Figure 6.7*.

   Ensure the center of the circle is positioned at the origin of the coordinate system.

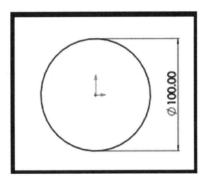

Figure 6.7 – Reference circle for the helix (100 mm diameter)

5. Launch the **Helix/Spiral** property manager from the main menu via **Insert →** **Curve → Helix/Spiral**.

6. Within the **Helix/Spiral** property manager that appears, create the helix guide curve with the details shown in *Figure 6.8*:

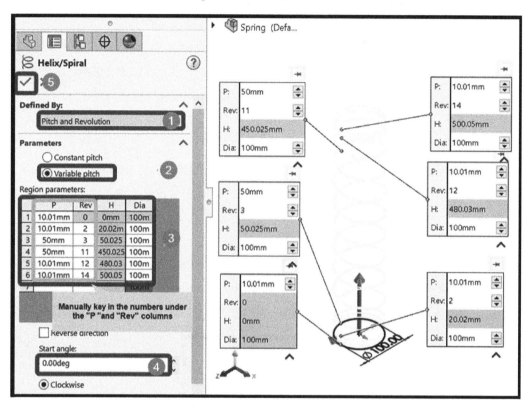

Figure 6.8 – Completing the helix profile

Notice that we have used the **Variable pitch** option (under **Parameters**) in *Figure 6.8*. You will also notice the preview of the helix. Notice that the first two coils and the last two coils will act as inactive coils.

With the completion of the helix profile, we will now add the cross-section of the spring wire.

Creating a new plane and the cross-section of the spring wire

The cross-section of the wire is 10 mm. However, to create this cross-section, we need a new plane. For this purpose, follow the steps listed next to create the new plane for the cross-section:

1. Navigate to the **Features** tab.

2. Click on **Reference Geometry**, and then select **Plane** (*Figure 6.9a*).

    From the **Plane** property manager that appears (after completing *step 2*), complete the creation of the plane by following the next steps.

3. Click inside the **First Reference** box (labeled 1 in *Figure 6.9b*), then navigate to the graphics window and pick the helix curve.

4. Click inside the **Second Reference** box (labeled 2), then navigate to the graphics window and pick the starting point of the helix curve.

5. Click **OK**.

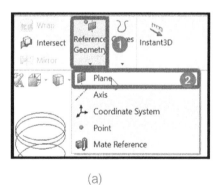
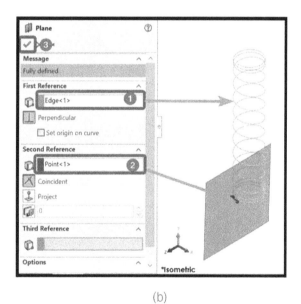

(a)   (b)

Figure 6.9 – Creating a reference plane for the cross-section

Steps 1–5 will create the plane. With the plane created, follow the steps listed next to create the 10 mm diameter cross-section.

6. With the new plane still active, use the **Circle** command to complete the sketching of the circle as shown in *Figure 6.10*:

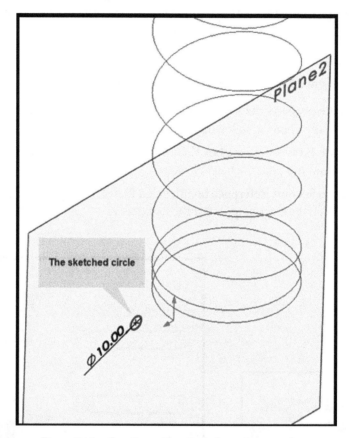

Figure 6.10 – Creating a 10 mm circle on the new plane

We now need to position the center point of the circle at the starting point of the helix. We will do this by adding a pierce relation between the sketched circle and the helix curve.

7. For the **Pierce** relation command to appear as shown in *Figure 6.11a*, hold down the *Ctrl* key, and then select the center of the circle. While still holding down the *Ctrl* key, click near the starting point of the helix (*don't aim for the starting point*). The pierce property manager should appear.

8. Click on the **Pierce** relations command (labeled 3), and then click **OK**.

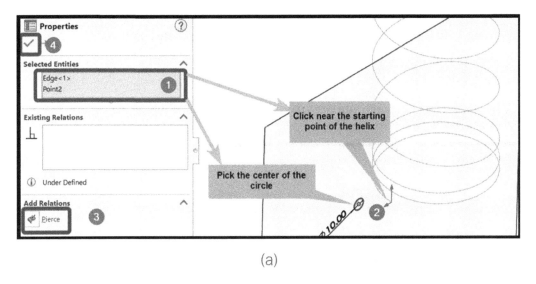

(a)

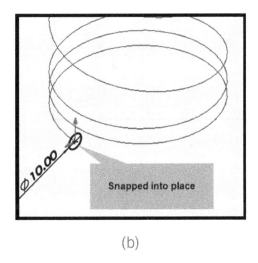

(b)

Figure 6.11 – (a) Applying the pierce relation; (b) illustration of the pierced circle/curve

With the cross-section snapped in place, we are now set to extrude the profile along the curve.

For this purpose, follow the steps listed next to use the **Swept Boss/Base** command to complete the extrusion:

1. Navigate to the **Features** tab and select the **Sweep Boss/Base** command.

2. Under the **Profile and Path** section, select the options indicated in *Figure 6.12a*.

3. Click **OK**.

    By completing the preceding steps, you will get a model of the spring as shown in *Figure 6.12b*:

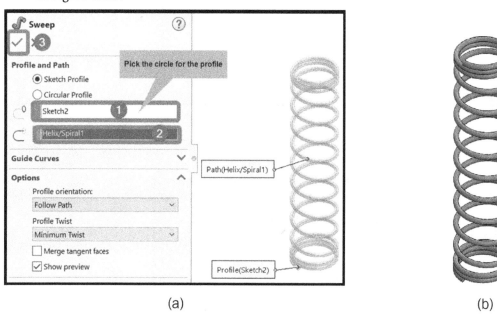

(a)   (b)

Figure 6.12 – Options for creating the sweep profile

We have now completed the basic model of the helical spring. But we are still left with three more details (somewhat minor but important) about our model of the spring sub-assembly. These are as follows:

*   *The end treatment*: This is what will give us the ground/flat ends. This eases the positioning of the spring as it interacts with other members of the assembly.

*   *A reference axis*: This allows us to read the stress value across a section of the curve.

*   *The models of the spring hub/load platen*: This will allow us to *apply load and boundary conditions* indirectly on the spring.

> **Note about End Treatment**
>
> For a variety of applications, it is common to subject compression helical springs to some form of end treatment. You can read more about this from any book on machine element design, for instance, the book by *Mott [4]* or from websites on spring design such as `https://www.acxesspring.com` or `https://www.leespring.com`.

We will now briefly engage with the creation of the end treatment.

For brevity's sake, the next set of steps offers a brief guide for the completion of the model. To create the end treatment and the reference axis, we will need to generate a set of intersection curves, which is the next task.

Adding construction intersection curves

Follow the steps given next to create the intersection curves:

1. Move to the **Feature Manager** tree, then click to activate **Right Plane**.

2. Right-click on the right plane to choose **Normal To** (*Figure 6.13a*).

3. Navigate to the **Sketch** tab and then select **Intersection Curves** under the **Convert Entities** command.

4. When the **Intersection Curves** property manager comes up, follow the illustration in *Figure 6.13b* to complete the generation of the intersected curves:

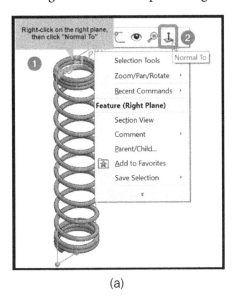 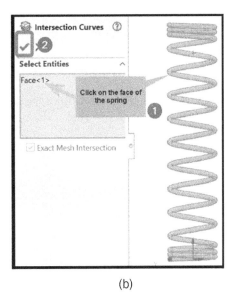

(a)   (b)

Figure 6.13 – Creating the intersection curves

After you are done with the **Intersection Curves** command, a projection of the cross-sections that intersect the right plane will be formed as shown in *Figure 6.14a*.

5. Select all the generated curves/circles (use the box selection approach) and convert them to construction sketches by ticking the **For construction** option (labeled 2 in *Figure 6.14b*).

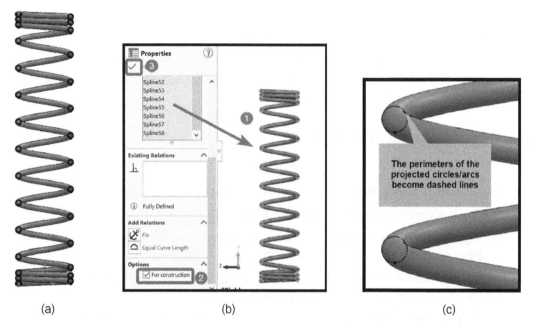

<table>
<tr><td>(a)</td><td>(b)</td><td>(c)</td></tr>
</table>

Figure 6.14 – Conversion of the intersection curves into construction sketches

After *step 5* is complete, you may zoom in on the converted circles/curves. You should notice their perimeters have turned into dashed lines as shown in *Figure 6.14c*.

We are now done with the creation of the construction intersection curves. How do we use the curves we've created? That is the focus of the next sub-section.

Adding the end treatments and reference axes

According to the problem statement, the spring is ground at the ends. We will use the **Extruded Cut** command to replicate the grounding of the spring's ends. The steps for doing this are as follows:

1. Using the tangent points on the converted arcs at the bottom and top of the spring, create the rectangles with the **Corner Rectangle** command as shown in *Figure 6.15a*.

2. Activate the **Extruded Cut** command from the **Features** tab. This will launch the **Cut-Extrude** property manager (*Figure 6.15b*).

3. Complete the options within the **Cut-Extrude** property manager (*Figure 6.15b*).

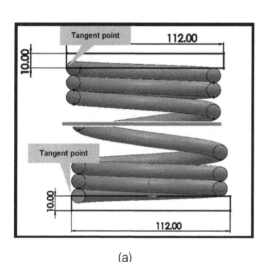

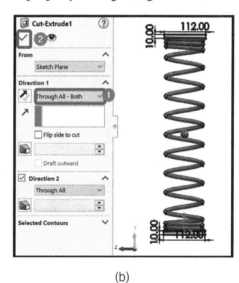

(a)   (b)

Figure 6.15 – Creating the end treatment

Next, we will add two reference axes. Each one will pass through any of the converted curves we created earlier. The steps to achieve the addition of the reference axes are given next:

1. Create a new sketch on the **Right Plane** (the same as *steps 1–5* in the sub-section *Adding construction intersection curves*).

2. Use the **Centerline** command to create two line sketches across any two of the circles (one is shown in *Figure 6.16a*). Ensure you exit the sketch mode after completing the line sketches.

3. Create two reference axes to coincide with the two lines created in *step 2* by selecting **Insert** from the main menu, and then **Reference Geometry → Axis**.

    Figure 6.16b shows the two reference axes created:

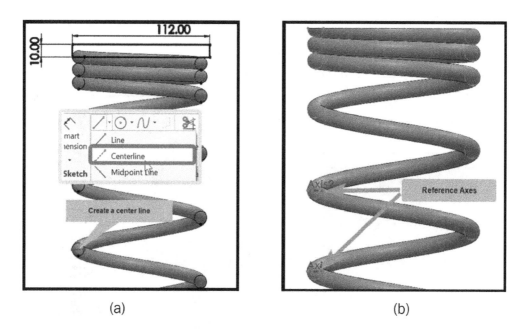

(a)   (b)

Figure 6.16 – Adding two reference axes

We have now completed the final status of the spring. One last thing we need to cater to is the creation of the spring hub base and the load platen. You may use the dimensions provided in *Figure 6.6(b and c)* to create these items as shown in *Figure 6.17a*. Notice that in *Figure 6.17b*, a reference point is placed at the center of the face to allow us to apply a point load later on. In any case, a complete model of the sub-assembly is available for download if you've any difficulty in completing the model shown in *Figure 6.17*. Primarily, the components of the sub-assembly are modeled in place (which means they were just placed on each other in one part file) with no interacting mates between them.

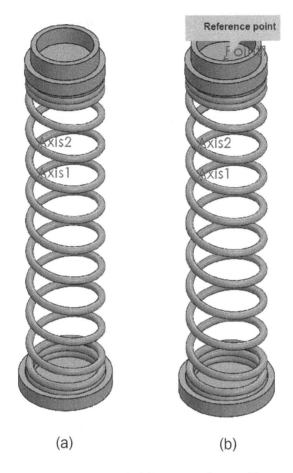

(a)   (b)

Figure 6.17 – Model of the spring sub-assembly

This brings us to the end of the modeling of the spring sub-assembly. We will now shift our focus to tackle the simulation tasks.

Part B: Creating the simulation study

The simulation workflow again comprises all the necessary pre-processing routines (the activation of the simulation study, assigning of material, application of fixtures, and the specification of loads).

We begin with the activation of the simulation environment.

Activating the Simulation tab and creating a new study

To activate the simulation commands, we follow the steps highlighted next:

1. Click on **SOLIDWORKS Add-Ins**.
2. Click on **SOLIDWORKS Simulation** to activate the **Simulation** tab.
3. With the **Simulation** tab active, create a new study by clicking on **New Study**.
4. Input a study name within the **Name** box, for example, `Spring analysis`.
5. Keep the **Static** analysis option.
6. Click **OK**.

As soon as we complete *steps 1–6*, you will see the simulation environment with the compartmentalized sections as shown in *Figure 6.18*:

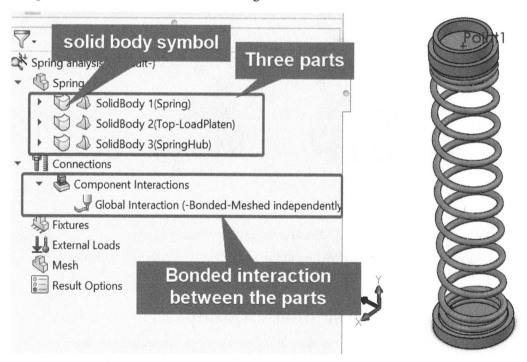

Figure 6.18 – Simulation property manager

As seen in *Figure 6.18*, we have three parts. One of the remarkable features of SOLIDWORKS Simulation is that, once you launch a simulation study, most parts are automatically treated as solids (unless converted otherwise). What this means is that we do not have to make any conversion of the three imported parts. Recall that in *Chapters 2, Analyses of Bars and Trusses* to *Chapter 4, Analyses of Torsionally Loaded Components*, we manually converted the bodies to trusses/beams. We also did a conversion to a shell in *Chapter 5, Analyses of Axisymmetric Bodies*. Since we want to use solid elements for our discretization, we will therefore accept this default behavior.

It's time to shift our attention to the application of materials.

Assigning the materials

For the purpose of simplification, the spring, the spring hub, and the load platen are taken to be made of ASTM316 stainless steel (this may not be the case in a real design).

Follow the steps given next to assign the material to the three bodies at once:

1. Right-click on the parts folder (named `Spring`) and choose **Apply Material to All Bodies** (*Figure 6.19*).

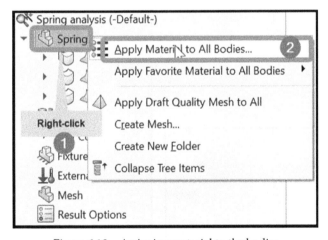

Figure 6.19 – Assigning material to the bodies

2. From the materials database that appears, expand the `Steel` folder.
3. Click on **AISI Type 316L stainless steel**.
4. Click **Apply** and **Close**.

The material assignment task is now completed. In the next sub-section, we will briefly take a look at the contact/interaction setting (before we deal with the application of the load and the specification of the fixture).

Examining the interaction setting

Since we have three parts that are interacting with each other, contacts must be defined between them. Now, given that the parts are designed to have faces that are touching each other, SOLIDWORKS has explicitly defined a contact for us.

To verify, let's examine this contact option:

1. Within the simulation study tree, expand the `Connections` folder, and then right-click on **Global Interaction** (*Figure 6.20a*).

2. From the **Component Interaction** property manager that appears, keep the options as shown in *Figure 6.20b*.

   As indicated in *Figure 6.20b*, a combination of **Bonded** interaction with the option to **Ensure common nodes between touching boundaries** has been chosen. The **Bonded** option selected here ensures the parts remain fully attached during the simulation (you don't want a situation where the spring runs away from the load platen).

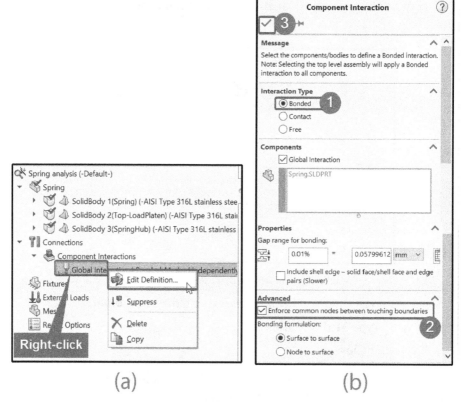

Figure 6.20 – Examining the interaction details

For more details about the other options, see *Chapter 4, Analyses of Torsionally Loaded Components*, in the sub-section *Scrutinizing the interaction settings*. There is so much on interaction/contact settings that we will come back to them in the second case study. We will also re-visit them in *Chapter 7, Analyses of Components with Mixed Elements*. Nevertheless, this wraps up the examination of the interaction settings.

It is now time to apply the fixture.

Applying the fixture at the base of the spring hub

Our boundary condition for this problem is simple: just a fixed support at the base of the sub-assembly. The steps to apply the fixture are as follows:

1. Right-click on **Fixtures**.

2. Pick **Fixed Geometry** from the context menu that appears (*Figure 6.21a*).

3. When the **Fixture** property manager appears, navigate to the graphics window and click on the bottom face of the sub-assembly as shown in *Figure 6.21b*.

4. Click **OK**.

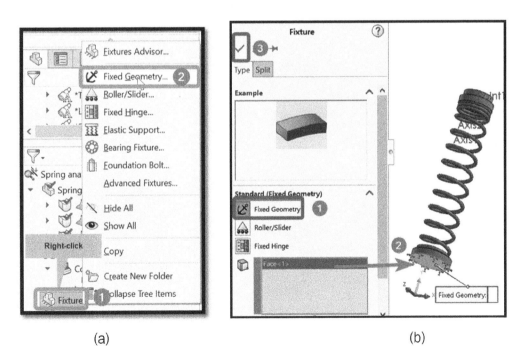

(a)   (b)

Figure 6.21 – Defining the fixed boundary condition

With the completion of the application of the fixed boundary condition, let's now move on to the application of the force at the top of the spring sub-assembly.

Applying the external load

To apply the force, follow the steps outlined next:

1. Under the simulation study tree, right-click on **External Loads** and select **Force**.

    Within the **Force/Torque** property manager that appears, execute the next actions.

2. Click inside the **Force** reference box (labeled 1 in *Figure 6.22a*) and navigate to the graphics window to click on the reference point at the center of the load platen.

3. Check the **Selected direction** option (labeled 3 in *Figure 6.22a*).

4. Click inside the direction reference box (labeled 4 in *Figure 6.22a*), and then navigate to the graphics window to choose **Right Plane**.

5. Under **Force**, input 255 in the box labeled 5 (Along Plane Dir 2) and ensure the unit system remains as **SI**.

6. Click **OK**.

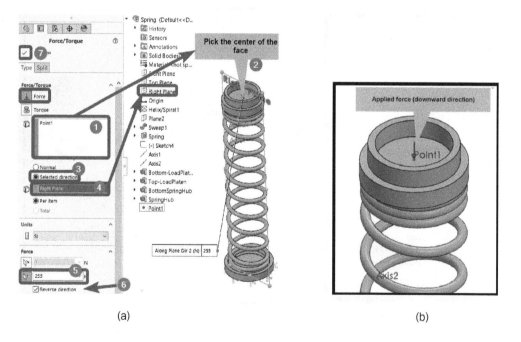

(a)   (b)

Figure 6.22 – Initiating the application of the point load

Figure 6.22b shows the arrow of the applied load. You will notice that we have applied a force of 255 N in *step 5*. This is derived from the conversion of the 25 kg mass to weight (250 N). This is then combined with the pre-load of 5 N coming from the weight of the load carrier.

At this point, we have covered the creation of a variable pitch helical compression spring. We have examined the state of the body as solid within the simulation environment. We have assigned a material, applied the necessary fixture, and applied the point load to the spring sub-assembly. We are therefore set for the meshing task and running of the analysis, which is what happens in the next section.

Part C: Meshing and post-processing of results

This sub-section is devoted to meshing. As stated in *Chapter 5*, *Analyses of Axisymmetric Bodies*, when dealing with advanced elements, the mesh quality has a profound effect on the accuracy of our results. Furthermore, as we are dealing with a spring in this analysis, it is worth exploring that a wrong choice of mesh type can also make or break a simulation workflow. So, in what follows, we will examine the difference between the meshes created by the *Standard Mesher* and the *Curvature-based Mesher*. In simple terms, the meshes created by the former are called standard meshes, while those of the latter are known as curvature-based meshes.

Examining standard and curvature-based meshes

Begin the meshing procedure by following the steps given next:

1. Right-click on **Mesh** within the simulation study tree and then select **Create Mesh**.

    The **Mesh** property manager appears, using which you need to make the following selections.

2. Under **Mesh Density**, drag the **Mesh Factor** control bar (labeled 1 in *Figure 6.23a*) to the right (indicating a fine mesh).

3. Under **Mesh Parameters**, go with the default choice of **Standard mesh**.

4. Click **OK**.

By completing *step 4*, you are likely to get a message indicating a failed mesh as shown in *Figure 6.23b*, although the other two components are meshed easily (indicated by the green grids).

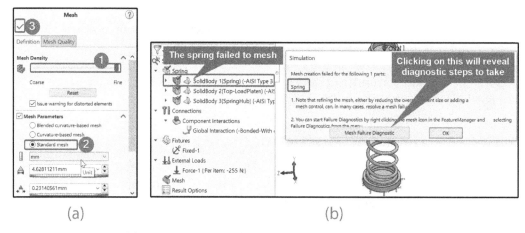

Figure 6.23 – (a) Options to create a fine standard mesh; (b) evidence of mesh failure

The message in *Figure 6.23b* suggests that the mesh failure is due to the spring. When you encounter this kind of failed meshing operation, clicking on the **Mesh Failure Diagnostic** button will take you to another window stipulating steps to take to correct the issue.

However, for this problem, we know that a major reason the mesh failed for the spring hinges on its extreme curvatures. The curved segments of the spring cannot be easily discretized using the standard mesh. We now have two options:

- The first option is to do mixed discretization. With this, the spring will be discretized using the curvature-based mesh approach and the other two components are discretized with the standard mesh.

- The second approach (and easy route) is to re-discretize all three components with the curvature-based mesh. Let's go with the second approach.

The steps to upgrade to a curvature-based mesh are briefly outlined here:

1. Right-click on the Mesh folder within the simulation study tree and select **Create Mesh.**

2.   Complete the options as indicated in *Figure 6.24a*:

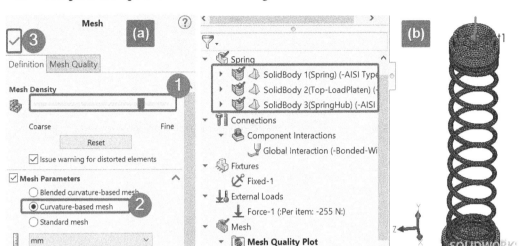

Figure 6.24 – (a) Options to obtain the curvature-based mesh; (b) the meshed solid bodies

As expected, completing the steps for the curvature-based mesh discretized all three bodies with no difficulty, as shown in *Figure 6.24b*. Why does the curvature-based mesher handle discretization easily? Well, the elements produced by this mesher are based on parabolic tetrahedral solid elements, which are more robust than the linear tetrahedral elements of the standard mesher.

Meanwhile, by virtue of the position of the mesh slider (labeled 1) in *Figure 6.24a*, you will notice that we are using a less than fine mesh. The mesh quality can be fine-tuned if our result is not satisfactory. But waiting to see the result before knowing whether the mesh is good for the analysis is not optimal. The trick is to use the mesh detail feature of SOLIDWORKS Simulation.

Therefore, let's take a look at the mesh detail, which we have not done in any of the previous chapters. To get the mesh detail, simply right-click on the Mesh folder and then select **Details** (*Figure 6.25a*). Doing so will reveal the features of the mesh shown in *Figure 6.25b*:

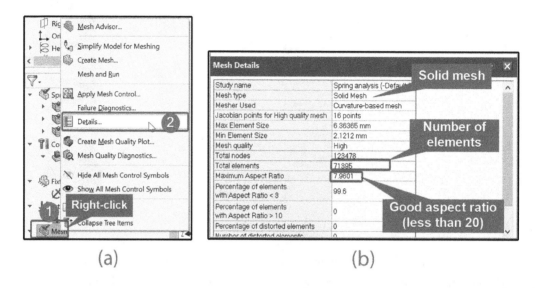

Figure 6.25 – Examining the mesh details

What can we garner from *Figure 6.25b*? We can get the following details:

- We have generated a set of solid elements (referred to as **Solid Mesh** in the figure).

- We have created the meshes using the curvature-based mesher.

- We have generated a total number of elements that equals *71,395* and created a number of nodes that equals *123,478*.

- We managed to have the **Maximum Aspect Ratio** for the element as *7.9601*.

A good rule of thumb to know whether you have a relatively good mesh is to ensure the maximum aspect ratio is not more than *20*. Luckily, this condition is satisfied by our mesh. For complex models, it is not uncommon to run into a situation in which the aspect ratio of some elements will be more than 20. In such scenarios, you will need to use **Mesh Control** to refine the discretization in the region where this happens. We will use **Mesh Control** in case study 2 later in this chapter.

Having come this far, we are ready to run the analysis and retrieve the results that we want.

Run the analysis

Follow the steps given next to accomplish the running operation:

1. Right-click on the study name and then select **Run** (*Figure 6.26a*).

2. You may get a suggestion about allowing the use of the **Large Displacement option**. Select **No** (*Figure 6.26b*).

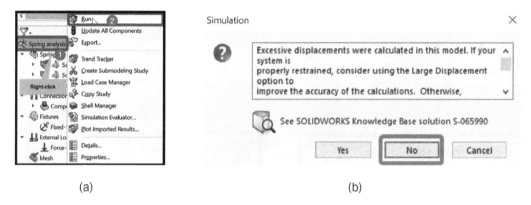

(a)   (b)

Figure 6.26 – (a) Initiating the running of the analysis; (b) suggestion to use the Large Displacement option

> **Note**
> Note that although we are neglecting the warning about excessive displacement here, the warning is telling us something about the design. Excessive deformation for a linear elastic material could lead to instability and should be checked in a normal design task. This warning will appear mostly when the maximum displacement/characteristic length of the structure being analyzed exceeds 10%.

Once the running process is complete, we will then be able to retrieve the desired results. Primarily, we are interested in the following:

- Determining the deflection experienced by the spring when the total load of 255 N is applied to it

- Finding the maximum shear stress developed in the spring

- Finding out how well the shear stress computed by the simulation compares with theoretical expressions

Let's commence with the retrieval of the deflection result.

Obtaining the deflection of the spring

We are interested in the displacement of the spring in the vertical direction (that is, along the Y-axis). To obtain this value, carry out the following steps:

1. Right-click on the `Results` folder and then select **Define Displacement Plot**.

2. We will now update some of the options within the **Displacement Plot** property manager that appears.

3. Under the **Definition** tab, navigate to **Display** and keep the options as shown in *Figure 6.27a*.

4. Now, move to the **Chart Options** tab, and then, under **Position/Format**, change the number format to **floating,** as depicted in *Figure 6.27b*.

5. Click **OK**.

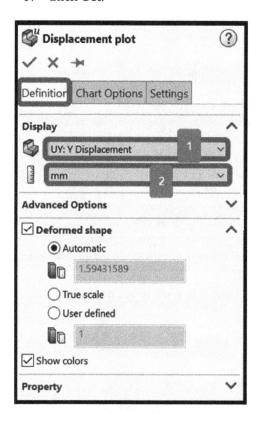

(a)

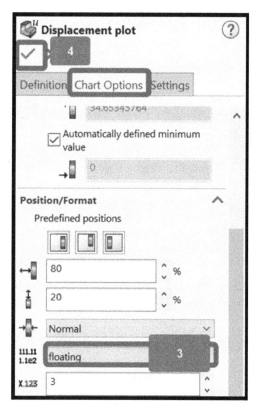

(b)

Figure 6.27 – Specifying options for the displacement plot

By completing *steps 1–5*, the graphics window will be updated with the plot of the vertical displacement, as shown in *Figure 6.28*:

Figure 6.28 – Distribution of the displacement across the three bodies

Figure 6.28 indicates that the maximum downward deflection happens to be *-34.5 mm*, which is at the upper region of the assembly, while the lower region experiences negligible deflection (which makes sense because it is fixed). Meanwhile, note that the maximum value is for the whole sub-assembly. Clearly, we will want to know the deformation experienced by the spring. For this purpose, we will fall back on the use of the **Probe** tool, as outlined next.

Using a Probe with the displacement

Follow these steps to retrieve the average value of the displacement on the spring:

1. Right-click on **Displacement2 (-Y disp-)**.

2. Select **Probe** from the resulting menu.

3. Keep the options in the **Probe Result** property manager as depicted in *Figure 6.29*.

4. Navigate to the graphic window and click on four points on the face of the spring in the upper region (see *Figure 6.2*).

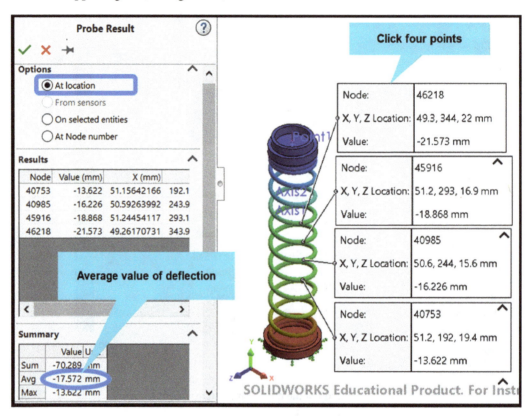

Figure 6.29 – Options for Probe Result and the average displacement value

You will notice a table summarizing the displacement values in the lower-left corner of the **Probe Result** window, as shown in *Figure 6.29*. The **Summary** table indicates that an average value of *-17.572 mm* is experienced by the spring.

But what can we say about the accuracy of this number? As is often the case, we can borrow some knowledge from the theories of the design of machine elements. For instance, for a spring with a wire diameter (d = 10 mm), a mean diameter (D = 90 mm), number of active coils (N_a = 10), shear modulus (G = 82 x 10^9 Pa), and an applied force (F), the deflection (y) can be predicted from equation 6-1 (see *Further reading [4]*):

$$y = \frac{F}{k}$$

(6-1)

Where k is the spring rate defined by:

$$k = \frac{d^4 G}{8D^3 N_a} = \frac{(0.01^4) \times (82 \times 10^9)}{8 \times (0.1^3) \times 10} = 14060 \frac{N}{m} \, or \, 14.06 \frac{N}{mm}$$

(6-2)

With *k* found (in equation 6-2), and F = 255 N, then y = 255/14.06 = 18.13 mm. As you can see, this value only differs by less than 1% from what we got from our SOLIDWORKS simulation. This brings us to the end of the discussion on the deflection of the spring.

We will now obtain the shear stress.

Obtaining the shear stress across the cross-section

Take the actions listed next to obtain the shear stress:

1. Right-click on the Results folder and then select **Define Stress Plot**.

   Now let's modify the options within the **Stress plot** property manager (summarized in *Figure 6.30*) that appears.

2. Under the **Definition** tab, navigate to **Display** and select **TXY: Shear in Y Dir**.

3. Within the unit box (labeled 2 in *Figure 6.30*), change the unit to **N/mm^2 (MPa)**.

4. Under **Advanced Options**, check the box labeled 3 (**Show plot only on selected entities**).

5. Under **Advanced Options**, click on the body symbol (labeled 4) and then navigate to the graphics window to click on the spring.

6. Under **Advanced Options**, click inside the box labeled 6 and then navigate to the graphics window to select one of the reference axes (you may repeat the step for the other reference axis).

7. Click **OK**.

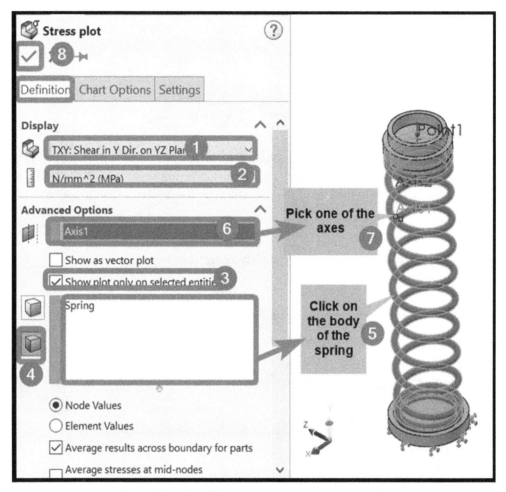

Figure 6.30 – Specifying options for the shear stress plot

By completing the preceding steps, you should obtain a plot similar to *Figure 6.31*:

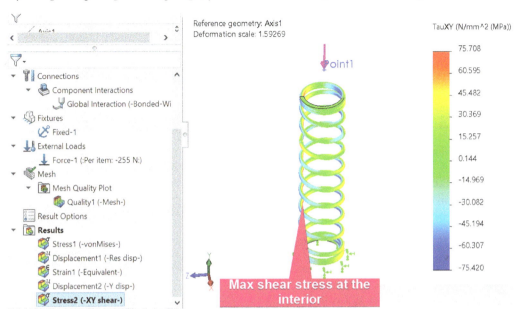

Figure 6.31 – Variation of the shear stress

On the one hand, it can be seen from *Figure 6.31* that the maximum shear stress happens to be roughly *75.708 MPa*. On the other hand, the maximum positive shear stress happens in the interior of the spring.

We will now compare the shear stress from the simulation with the theoretical equation for shear stress in a spring, which was originally derived by Wahl [5]:

$$\tau_{max} = K_s\left(\frac{8FD}{\pi d^3}\right) + \frac{4F}{\pi d^2}$$

(6-3)

where $K_s = \dfrac{2C+1}{2C}$, and it referred to the shear correction factor. Besides $C = D/d$ (the spring index), the other parameters are as defined in the earlier equations. Substituting the values for all the parameters into the equation 6-3 yields *71.42 MPa*.

Again, the theoretical value of the shear stress and that obtained from the simulation are very close, confirming the accuracy of the simulation setup. Recall that we have used a less than fine mesh, so the gap is likely to close with further refinement of the mesh. You are encouraged to try this to confirm.

This concludes the simulation of the first case study. This specific case study has taken us on a journey that sees us modeling a helical spring, preparing it for downstream simulation tasks. Furthermore, we have obtained, examined, and validated the displacement and shear results obtained for the spring.

Before we wrap up the chapter, we will go through another case study in the next section.

Analysis of spur gears

The central objective of this section is to analyze the spur gears assembly. We will focus narrowly on the qualitative assessment of the stress arising from the interaction of the gears' teeth. Among other things, this case study will enable us to see the application of a local interaction contact and the use of mesh control.

Problem statement – case study 2

Figure 6.32a shows the assembly of a gear-shaft system for a power transmission train. The gears are made of plain carbon steel. The purpose of this case study is to examine the stress that develops due to the contact between the teeth of the gears when gear A rotates by 1.15 degrees, while gear B is retrained.

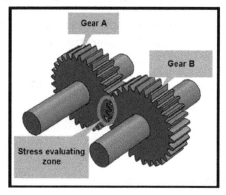

(a) Basic model of the gear-shaft assembly   (b) Model of the gears with split lines

Figure 6.32 – A gear-shaft assembly

To address the simulation task concisely, two models of the assembly are provided for this problem. Download the `chapter 6` folder from the book's GitHub repository. Inside the folder, you will see the first assembly file (named `Gear-Assembly`), which is the basic model shown in *Figure 6.32a*. The second assembly file is named `Gear-Assembly-Split`, and is the one shown in *Figure 6.32b*, where split lines are added to facilitate mesh control and a stress plot. In what follows, the second assembly file will be used for the analysis.

Launching SOLIDWORKS and opening the gear assembly file

Here, we start by opening the assembly file in SOLIDWORKS (**File → Open → Gear-Assembly-Split**). Ensure the unit is set to the **MMGS** unit system.

Once the assembly is opened, it is good to check any interference between the parts.

Follow the steps given next to check the interference of the assembly:

1. Navigate to the **Evaluate** tab, and click on **Interference Detection** (*Figure 6.33a*).

   This launches the **Interference Detection** manager, inside which we take the steps that come next.

2. Within the **Selected Components** box (labeled 1 in *Figure 6.33b*), click on the assembly in the graphics window.

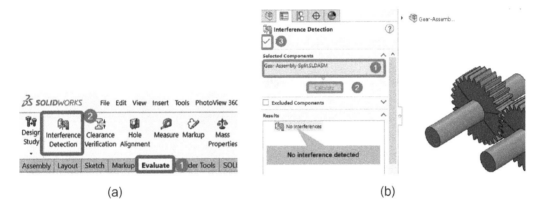

| (a) | (b) |

Figure 6.33 – Checking interference

3. Next, click on the **Calculate** button, labeled 2 in *Figure 6.33b* (notice that this will be grayed out after you click it). The outcome of the interference calculation will be revealed in the **Results** box.

4. Click **OK**.

As you can see in *Figure 6.33b*, there is no interference in the assembly. Therefore, we will move to the simulation phase.

Creating the simulation study for the gear and simplifying the assembly

For this purpose, follow the steps given next:

1. Using the **Simulation** tab, launch a **New Study**.

2. From the **Study** property manager that opens, provide a name for the study, for example, Gear analysis.

3. Click **OK**.

After completing *steps 1–3*, the simulation environment is launched. You will then notice the four bodies contained in the assembly, as shown in *Figure 6.34a*. For this analysis, we are interested in the gears. Since the shafts are not of interest for this simulation, they will be removed using one of the useful features of the simulation environment (**Exclude from Analysis**).

To achieve the exclusion of the shafts from the analysis, carry out the following steps:

1. Select the two shafts and right-click within the selected region, as shown in *Figure 6.34b*.

2. From the menu that appears, choose **Exclude from Analysis**.

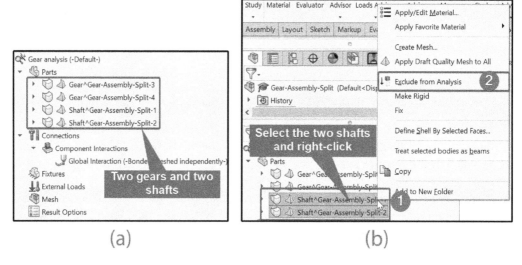

Figure 6.34 – (a) Four bodies in the assembly; (b) excluding the shafts from the analysis

The preceding steps will smartly prevent the two shafts from being considered in our analysis. This has the advantage of reducing the size of the components that will be meshed and the duration of our analysis. You won't always have to do this, but it is a useful tool to be aware of.

Having achieved the removal of the shafts, our next task is to apply materials.

Assigning materials to the gears

The material for the gears is plain carbon steel. Specify the material by following the steps given next to assign the material to the three bodies at once:

1. Right-click on the two gears and choose **Apply/Edit Material** (*Figure 6.35*).

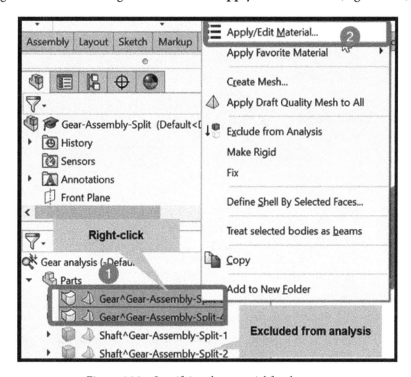

Figure 6.35 – Specifying the material for the gears

2. From the materials database that appears, expand the `Steel` folder.

3. Click on **Plain Carbon Steel**.

4. Click **Apply** and **Close**.

Our next task is setting up the interaction condition. This is crucial for getting a realistic result for the gear analysis.

Setting up the interaction condition for the gears

As mentioned in the sub-section *Examining the interaction setting* (for case study 1), the default behavior of SOLIDWORKS is to define a bonded interaction for bodies/components that are in contact. Now, since the gears will be in contact but not bonded, for this case study, the interaction condition needs to be modified.

To set up the right contact, carry out the following steps:

1. Expand the `Connections` folder and then delete the **Global Interaction** (*Figure 6.36*).

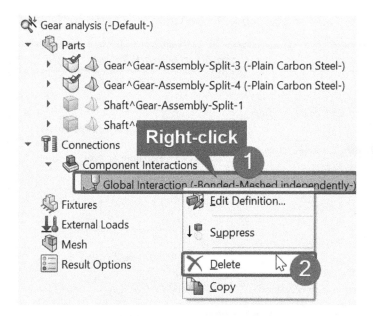

Figure 6.36 – Deleting the bonded contact set

2. Next, right-click on the `Connections` folder again and choose **Local Interactions**.

   This launches the **Local Interactions** manager, inside which we take the steps that come next.

3. Within the **Local Interactions** window, under **Interaction**, select **Manually select local interactions** (labeled 1 in *Figure 6.37*).

4. Under **Type**, select **Contact** (labeled 2 in *Figure 6.37*).

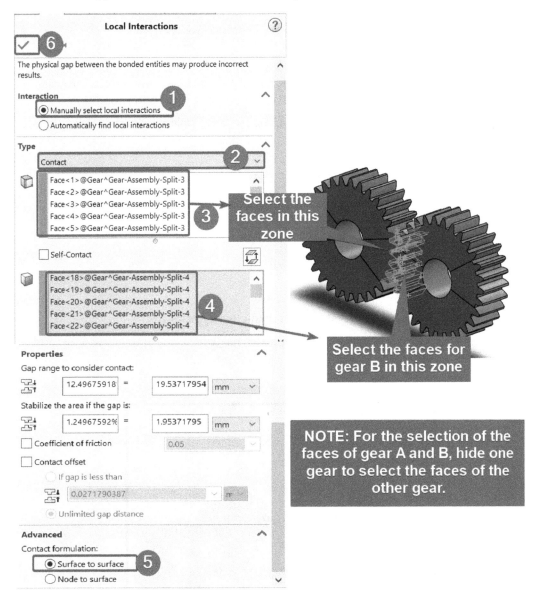

Figure 6.37 – Setting up a local contact interaction

5. Under the box for **Set 1** (labeled 3 in *Figure 6.37*), navigate to the graphics window and click on the contacting faces in the split zone of gear A.

6. Under the box for **Set 2** (labeled 4), navigate to the graphics window and click on the contacting faces on gear B (they should be opposite those of gear A).

7. Under **Advanced**, keep the option as **Surface to surface** (labeled 5).

8. Click **OK**.

Completing *steps 1–8* facilitates the right condition for the interaction between the gears. A bonded interaction, on the other hand, will create a wrong interaction condition between the gears since it will treat the faces as if they were welded together. We are now ready to shift our focus to the fixture.

Applying the fixture to the gears

Each of the two gears needs to be properly supported to behave the way we want. We will specify that gear B is fixed, while gear A can rotate around the Z-axis (based on the problem statement).

The steps to apply the condition for gear B go like this:

1. Right-click on **Fixtures**.

2. Pick **Fixed Geometry** from the context menu that appears.

3. When the **Fixture** property manager appears, navigate to the graphics window and click on the inner face of gear A, as shown in *Figure 6.38*.

4. Click **OK**.

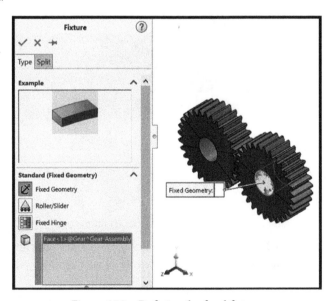

Figure 6.38 – Defining the fixed fixture

We will now carry out the steps to apply the right boundary condition for gear A:

1. Right-click on **Fixtures**.

2. Pick **Advanced Fixtures** from the context menu that appears.

3. When the **Fixture** property manager appears, click the **On Cylindrical Faces** option (labeled 1), then navigate to the graphics window and click on the inner face of gear A, as shown in *Figure 6.39*.

4. Under **Translations**, specify the translation options as shown in the boxes labeled 3–5.

5. Click **OK**.

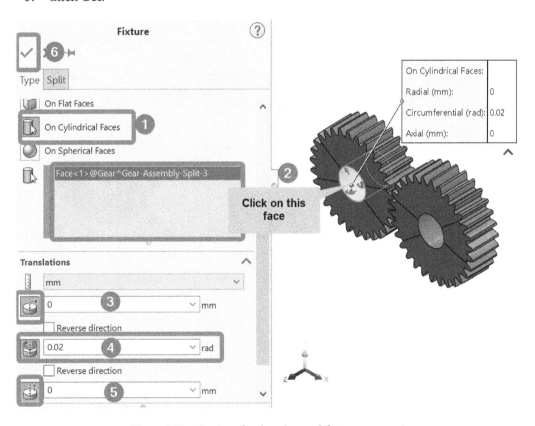

Figure 6.39 – Options for the advanced fixture on gear A

Notice that for the circumferential rotation (labeled 4 in *Figure 6.39*), we have entered the value 0.02 radian. This is equivalent to the value of 1.15 degrees specified in the problem statement.

With the completion of the applications of the boundary conditions, we are ready for the meshing phase, which is explored in the next sub-section.

The meshing of the gears

For the meshing operation, we will first create a base mesh (using the curvature-based mesh) and then we will refine the mesh in the contact region. To achieve our aim, take the following steps:

1. Right-click on **Mesh** and then select **Create Mesh**.

    Complete the options as indicated in *Figure 6.40a*. Notice that we are using the curvature-based mesh. The meshed gears are shown in *Figure 6.40b*:

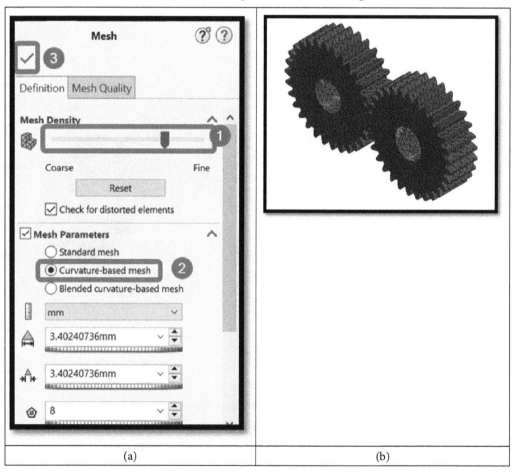

(a)	(b)

Figure 6.40 – (a) Options to select the curvature-based mesh; (b) the meshed solid gears

The mesh created in *step 1* is applied to the whole body of the two gears. To have a more refined mesh around the contact region, we will deploy the **Mesh Control** command as follows.

2. Right-click again on the Mesh folder in the simulation study tree and then select **Apply Mesh Control** (*Figure 6.41a*).

3. When the **Mesh Control** property manager appears, under **Selected Entities**, click inside the reference entities box (labeled 1). Navigate to the graphics window and click on the split faces as shown in *Figure 6.41b*.

4. Drag the **Mesh Density** slider (labeled 3) to the rightmost end (fine density).

5. Click on **Create Mesh**.

6. Click **OK**.

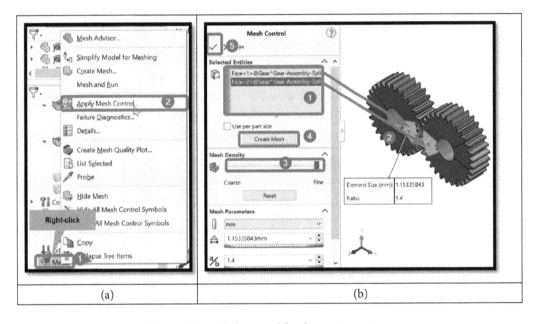

| (a) | (b) |

Figure 6.41 – Mesh control for the contact regions

The difference between the area we have applied mesh control to and the other areas of the gears can be seen in *Figure 6.42*:

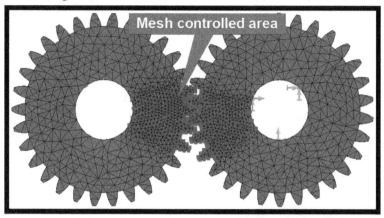

Figure 6.42 – Refined mesh in the contact zone

You can check the mesh details to examine the quality of the mesh (as we did in the previous sub-section *Examining standard and curvature-based meshes*). As seen in *Figure 6.43*, the mesh quality is satisfactory. We have a **Maximum Aspect Ratio** of around 9. Further, we also see evidence that we are indeed using a set of solid elements (a total number of 44225).

Mesh Details	
Study name	Gear analysis (-Default-)
Mesh type	Solid Mesh
Mesher Used	Curvature-based mesh
Jacobian points for High quality mesh	16 points
Mesh Control	Defined
Max Element Size	3.40241 mm
Min Element Size	3.40241 mm
Mesh quality	High
Total nodes	68971
Total elements	44225
Maximum Aspect Ratio	8.8497
Percentage of elements with Aspect Ratio < 3	96.8
Percentage of elements with Aspect Ratio > 10	0
Percentage of distorted elements	0

Figure 6.43 – Examining the mesh details for the gears

Now that we have gotten the meshing out of the way and conducted an examination of the generated mesh, let's run and retrieve the results.

Running and obtaining the results

As you know, once the other sub-sections are all properly set up, running the analysis is done in a single step. That is, right-click on the study name (Gear analysis) and then select **Run,** as shown in *Figure 6.44*:

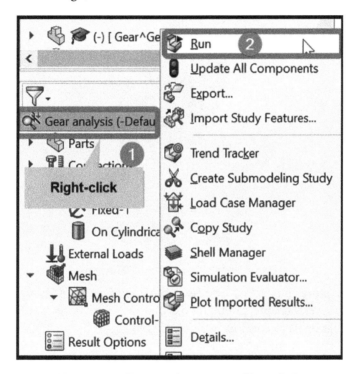

Figure 6.44 – Running the gear assembly analysis

Once the running of the simulation is complete, we will have the set of default results as usual. Indeed, *Figure 6.45* depicts the plot of the **von Mises stress** from among the default results. From the figure, you will notice that a large portion of the two gears experience almost negligible stress from the applied load (mostly blue). However, the contact region is highly stressed.

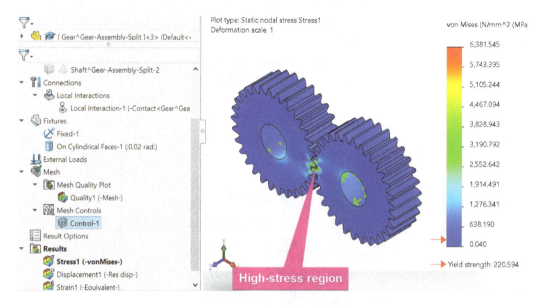

Figure 6.45 – Distribution of the von Mises stress

The design of gears to prevent excessive stress as shown in *Figure 6.45* is a specialized topic that is covered over several chapters in textbooks on the design of machine elements. Our interest in this simulation is to confirm two points *[4]*:

- The largest contact stress will happen at the mating surface of the teeth.

- The largest bending stress will take place at the root of the teeth in contact due to stress concentration.

We'll move on to investigate the stress at the contact region qualitatively. To confirm the first point, carry out the following steps:

1. Right-click on the `Results` folder and then select **Define Stress Plot**.

    Next, adjust the options within the **Stress plot** property manager that appears as follows:

2. Under the **Definition** tab, below the **Display** option, select **Contact Pressure** (*Figure 6.46a*).

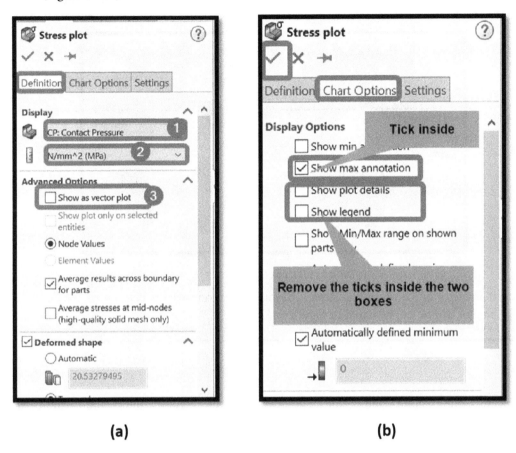

(a)   (b)

Figure 6.46 – Options for the contact stress

3. Under the **Definition** tab, within the unit box, change the unit to **N/mm^2 (MPa)**

4. Under the **Definition** tab, below **Advanced Options**, remove the checkmark inside the box labeled 3 in *Figure 6.46a* (**Show as vector plot**).

5. Under the **Chart Options** tab, select the options as shown in *Figure 6.46b*.

6. Click **OK**.

Once *steps 1–6* are complete, the environment is updated with the plot of the contact stress as indicated in *Figure 6.47*:

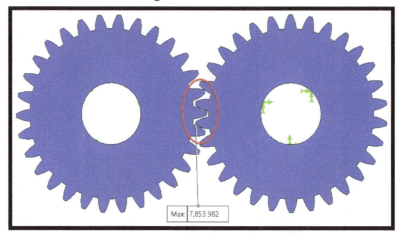

Figure 6.47 – Visualization of the position of the maximum contact stress

Now, if, in *step 4* of the preceding *steps 1–6*, you did not remove the checkmark inside the box, the plot will be as presented in *Figure 6.48*:

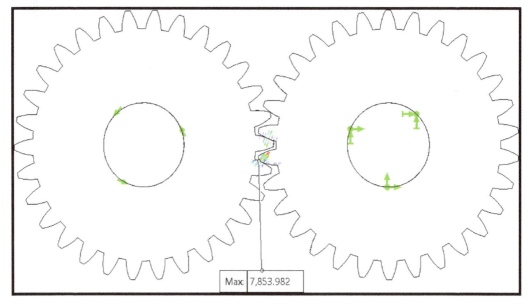

Figure 6.48 – Visualization of the contact stress as a vector plot

Both visualizations in *Figure 6.47* and *Figure 6.48* confirm that the maximum contact stress indeed happens at the mating surface.

The last result we will look at is the first principal stress. We are using the principal stress as a proxy for bending stress. For this purpose, take the steps outlined next to obtain the principal stress:

1. Right-click on the Results folder and then select **Define Stress Plot**.

2. Within the **Stress plot** property manager, under the **Definition** tab, navigate to **Display** and select **P1: 1ˢᵗ Principal Stress** (*Figure 6.49a*).

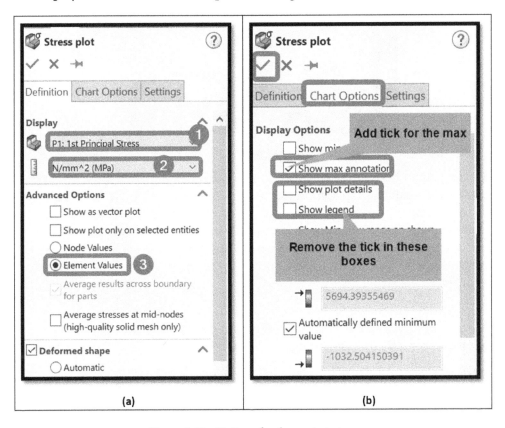

Figure 6.49 – Options for the contact stress

3. Under the **Definition** tab, within the unit box, change the unit to **N/mm^2 (MPa)**.

4. Under the **Definition** tab, below **Advanced Options**, check the box labeled 3 in *Figure 6.49a* (**Element Values**).

5. Under the **Chart Options** tab, select the options as shown in *Figure 6.49b*.

6. Click **OK**.

With the completion of *steps 1–6*, the environment is again updated with the plot of the principal stress, as shown in *Figure 6.50*:

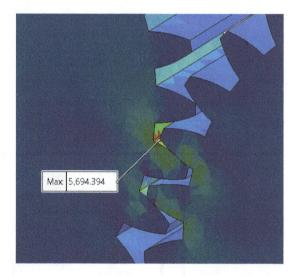

Figure 6.50 – Visualization of the first principal stress at the root of the teeth

Again, it can be seen from *Figure 6.50* that the maximum bending stress indeed happens at the root of the tooth that experiences the maximum contact stress. We have only focused on qualitative assessments in the last two results to bring home the concepts. For some other variations of this case study, the numerical values of the stress need to be checked for failure analysis and so on. Besides, it is also possible to employ symmetry for the gear problem (for instance, by using a quarter of each gear). We can also do further mesh refinement and so on.

This wraps up our exploration of the two case studies. Over the course of the two case studies, we've explored some additional features of SOLIDWORKS Simulation. For instance, we demonstrated the following:

- How to employ curvature-based meshes for highly curved members
- How to simplify solid bodies for analysis with the **Exclude from Analysis** command
- How to examine mesh details to confirm the quality
- How to employ mesh control to refine some selected regions of components
- How to set up a contact interaction relationship
- How to obtain and visualize contact pressure/stress

All these concepts are valuable for the simulation of many other types of components and you can extend the idea further in your various design tasks.

Summary

This chapter set out to explore the use of solid elements. It began with an overview of components that are suitable to be analyzed with solid elements and then shifted gear to the analysis of two case studies. The first case study focused on a spring sub-assembly. This was then followed by the second case study on the static analysis of a gear sub-assembly. Along the way, we leveraged on many features of SOLIDWORKS Simulation that are useful for the analysis of moderately complex entities using solid elements.

Up to now, we have dealt with the analysis of components using a single family of elements in each of *Chapters 2, Analyses of Bars and Trusses – Chapter 6, Analyses of Components with Solid Elements*. In the next chapter, you will learn about analyzing components with mixed elements.

Exercises

1. A variable pitch spring is to be made from a music (ASTM A228) wire with a diameter of 20 mm and other data as shown here:

   Total number of active coils = 10

   Number of inactive coils = 4

   Free length = 500 mm

   Outside diameter = 120 mm

   The spring is ground at the ends. Compute the deflection experienced by the spring when a force of 300 N is applied at its top.

2. Repeat case study 2 for cases where: (a) the friction coefficient is set to 0.02 by evaluating the effect of the change of the coefficient of friction on the stress values; and (b) the default bonded interaction is retained with no manual definition of the "contact" interaction. How does this affect the stress distribution?

Further reading

- [1] *Fundamentals of Finite Element Analysis: Linear Finite Element Analysis, I. Koutromanos, John Wiley & Sons, 2018.*

- [2] *Introduction to the Finite Element Method 4E, J. N. Reddy, McGraw-Hill Education, 2018.*

- [3] *Finite element analysis of a deployable space structure, D. V. Hutton, Edited by Dr. B. F. Barfield, Professor of Mechanical Engineering, the University of Alabama Tuscaloosa, AL, 1982.*

- [4] *Machine Elements in Mechanical Design, R. L. Mott, Pearson/Prentice Hall, 2004.*

- [5] *Mechanical springs. Penton Publishing Company, A. M. Wahl, 1944.*

- [6] *Introduction to Solid Modeling Using SolidWorks 2009, W. Howard and J. Musto, McGraw-Hill Companies Incorporated, 2009.*

- [7] Space Modeling with SolidWorks and NX, *J. Duhovnik, I. Demsar, and P. Drešar,* Springer International Publishing, 2014.

7

Analyses of Components with Mixed Elements

Thus far in this book, we have explored simulation examples in which components are discretized using a single family of elements. However, as can often happen with many complex practical design problems with diverse functional components, you will face situations where the goal will be to put these elements together in one single simulation task. Consider, say, the simulation of car frames, building frameworks, or plane fuselages with all the flooring details and connections. By taking advantage of putting different elements together, you can accelerate the simulation runtime or reduce the computational resources required to achieve your simulation objectives.

For the above reason, the elements you have seen in the previous chapters will make a re-appearance again in this chapter. We will examine two case studies, and our discussion will orient around the following topics:

- Analysis of three-dimensional components with mixed beam and shell elements
- Analysis of three-dimensional components with mixed shell and solid elements

Technical requirements

You will need to have access to the SOLIDWORKS software with a SOLIDWORKS Simulation license.

You can find the sample files of the models required for this chapter here: `https://github.com/PacktPublishing/Practical-Finite-Element-Simulations-with-SOLIDWORKS-2022/tree/main/Chapter07`.

Analysis of three-dimensional components with mixed beam and shell elements

This section details our first case study, which centers around an analysis that involves a mixture of **beam elements** and **shell elements**.

As you will recall, we pointed out the attributes of the beam element in *Chapter 3, Analyses of Beams and Frames*, while those of the shell elements were described in *Chapter 5, Analyses of Axisymmetric Bodies*. Although their dimensionality is different (one being a line-based element and the other being a triangular element), both of these elements as implemented in SOLIDWORKS Simulation have six **degrees of freedom** (**DOFs**) per node. Notably, these DOFs are the three translational displacements along the X, Y, and Z axes, and the three rotational displacements about these three axes. Putting these two elements together in the first case study, therefore, represents a good starting point in the exploration of analyses with mixed elements. Overall, the compatibility of the DOFs is one of the more important criteria to consider when employing mixed elements to speed up your simulation. If you want to dig deeper into other practical considerations for the use of mixed meshing, you may refer to *[1]*.

The premise for the case study is given next.

Problem statement – case study 1

Figure 7.1a shows an idealized assembly representing the basic structural elements of a two-story residential building *[2]* (see the *Further reading* section). The major dimensions of the building (height x width x length) are given by 6 m x 6 m x 18 m. The beams and columns supporting the building floors are made of W410x16 ASTM A36 structural steel. Each has the cross-section shown in *Figure 7.1b*. In addition, the 130 mm thick floor slabs are made of concrete (*E = 30 GPa, G = 12.71 GPa, Poisson's ratio = 0.18, ultimate compressive strength = 41 MPa, ultimate tensile strength = 5 MPa, and density = 2300 kg/m³*).

We aim to conduct a simplified static analysis of the building under a dead load arising from the self-weights of the structural members and a live load of 2500 N/m² on the floor slabs. Let's assume the value of the dead load complies with the local residential structural design guide *[3]*. From the analysis, we shall examine the maximum vertical displacement of the building under these loads and the bending stresses in the beams and the columns.

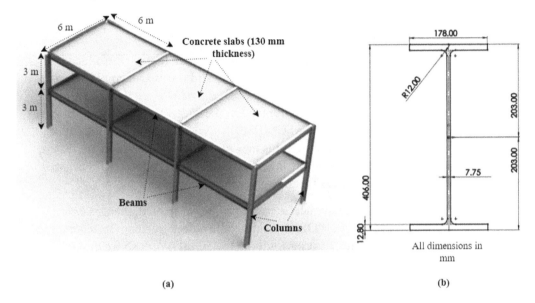

(a)   (b)

Figure 7.1 – The model of a partitioned two-story building

In what follows, we will address the problem in four major sections designated as *Part A: Reviewing the structural models*, *Part B: Creating the simulation study*, *Part C: Meshing and running*, and *Part D: Post-processing of results*.

Let's begin.

Part A: Reviewing the structural models

To get you started, the files of the models we need have been provided. This means you should download the Chapter 7 folder from the book's website to retrieve the models. Inside the folder, you will see the following items:

- A SOLIDWORKS part file named Two-story
- A SOLIDWORKS part file named Two-story-Solar
- A folder named Custom

On the one hand, the first and second items pertain to the full model of the two-story buildings for case studies 1 and 2, respectively. On the other hand, the third item is a folder that contains the file of the sketched profile shown in *Figure 7.1b*.

We will briefly go over the first and third items in the next sub-sections. The second item will be briefed when we get to our second hands-on.

Copying and reviewing the file in the custom profile

First, let's clarify why we need the `Custom` folder. You will recall that we introduced the weldment tool in *Chapter 2, Analyses of Bars and Trusses*, and it was also employed in *Chapter 3, Analyses of Beams and Frames*. In those chapters, we worked with the original weldment profiles from the SOLIDWORKS library. On the contrary, the cross-section shown in *Figure 7.1b* is not available in the SOLIDWORKS **weldment** database. The consequence of this is that we need to create this new profile. Once the profile is created, it should be saved in a separate folder to distinguish it from the SOLIDWORKS original files. This is the reason for having the `Custom` folder for this chapter.

Meanwhile, for SOLIDWORKS to have access to the profile we created, the `Custom` folder has to be placed in the SOLIDWORKS weldment profiles directory. To accomplish this, follow these steps to move the `Custom` folder to the SOLIDWORKS weldment profiles directory:

1. Navigate to the `Chapter 7` folder that you have downloaded.
2. Copy the folder named `Custom` (using the normal Windows file copying procedure).
3. Paste the folder into the SOLIDWORKS in-built weldment directory at the following address: `Drive:\Program Files\SOLIDWORKS Corp\ SOLIDWORKS\lang\english\weldment profiles)`.

    After pasting, you should see the `Custom` folder among the other directories as shown in *Figure 7.2*.

Figure 7.2 – The default and the custom directories

You should examine the content of the Custom folder to see that it has another sub-folder called Custom_ISO. In turn, this sub-folder contains the created profile (**W410 x 60**) as depicted in *Figure 7.3*.

Figure 7.3 – Custom folder for the created profile

You will require administrator rights on the computer you are using to be able to copy into the SOLIDWORKS in-built weldment directory.

> **Note on Weldment**
>
> We explored the default weldment profile in *Chapter 2, Analyses of Bars and Trusses*, in the sub-section, *Introducing the weldments tool*. Further, in that chapter, we examined the content of the ansi inch and iso folders shown in *Figure 7.2*. More importantly, for a brief description of how to create a custom profile such as W410 x 60, a pertinent resource is the official SOLIDWORKS help link: https://help.solidworks. com/2021/English/SolidWorks/sldworks/t_Weldments_ Creating_a_Custom_Profile.htm.

Note that *steps 1–3* implicitly assume that you have never created a custom weldment sub-directory before. However, if you already have a specific folder for your custom weldment profiles, you should only copy and paste the W410 x 60.SLDLFP file into that folder.

With the custom profile copied into the SOLIDWORKS library, we shall now take a look at the file of the building (that is, Two-story).

Reviewing the file of the building

To go over some of the features of the model of the building, follow these steps to bring the file up for examination:

1. Start up SOLIDWORKS.

2. Select **File → Open**, then open the part file named Two-story (which you downloaded).

3. Navigate to the **Feature Manager** design tree and examine the details as shown in *Figure 7.4*.

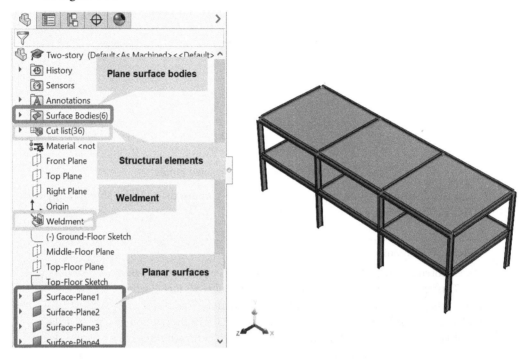

Figure 7.4 – Review of components

Figure 7.4 points to the fact that we have *6 surface bodies* to idealize the concrete slabs and *36 cut lists* to represent the structural elements (beams and columns). Further, you can see that we have used the planar surface tool (**Insert → Surface → Planar**) to create the surface bodies, while the structural elements have been created using the weldment tool.

With the files reviewed, we are now primed to tackle the simulation tasks.

Part B: Creating the simulation study

This section details some of the necessary preprocessing routines (the activation of the simulation study, assigning of material, application of fixtures, and specification of loads).

We begin with the activation of the simulation environment.

Activating the Simulation tab and creating a new study

To activate the simulation commands, we follow the steps highlighted next:

1. Under the **Command manager** tab, click on **SOLIDWORKS Add-Ins**, then click on **SOLIDWORKS Simulation** to activate the **Simulation** tab.

2. With the **Simulation** tab active, create a new study by clicking on **New Study**.

3. Input a study name within the **Name** box, for example, `Two-story analysis`.

4. Keep the **Static** analysis option.

5. Click **OK**.

In response to the completion of *steps 1–5*, the model is transformed in the simulation environment with the joints created at the connection points of the beams and the columns as shown in *Figure 7.5a*.

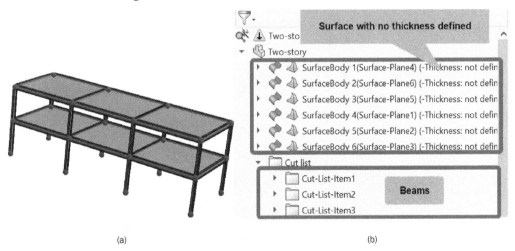

(a)   (b)

Figure 7.5 – Simulation property manager

Looking at the simulation study tree (partially depicted in *Figure 7.5b*), you will notice the 6 surface bodies and the `Cut-List` folders comprising the beams and columns. Coupled with the previous detail, you will observe that the thickness has not been defined for the surface bodies as highlighted in *Figure 7.5b*. So, our next focus is to define the thickness.

Assigning thickness

The surface bodies will be discretized using shell elements during the meshing process. For this reason, we need to define the thickness using the **Shell Definition** tool. Follow these steps to assign thickness:

1. Within the simulation study tree, select all 6 surface bodies, right-click within the selected area, then select **Edit Definition...** as shown in *Figure 7.6*.

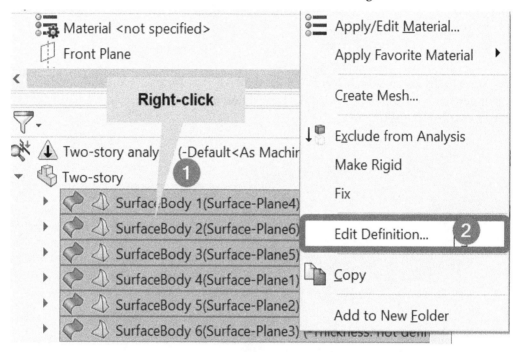

Figure 7.6 – Initiating the thickness definition for the surface bodies

Within the **Shell Definition** property manager that appears, perform the following actions.

2. Under the **Type** option, select **Thin** as shown in *Figure 7.7*.

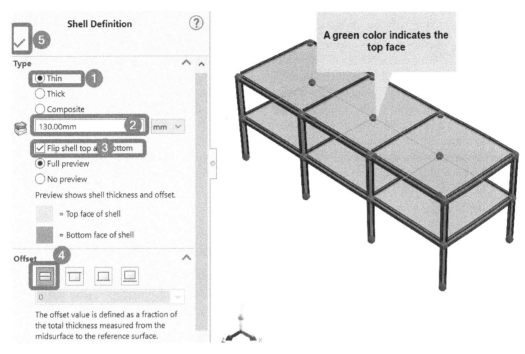

Figure 7.7 – Shell Definition property manager

3. Within the thickness value box (labeled 2 in *Figure 7.7*), key in 130 mm.

4. Check the box labeled 3 (**Flip shell top and bottom**).

5. Activate the **Full preview** option.

6. Under **Offset**, ensure the mid-surface option is selected.

7. Click **OK**.

This completes the specification of both the thickness of the surface bodies and the choice of a thin shell assumption. A quick remark on *step 4*. Essentially, we have used this step to distinguish which face of the shell elements the pressure load acts upon. This step is crucial anytime we create a surface body using the planar surface tool. The application of external loads or supports on the wrong side of shell elements may at times throw some unexpected behavior.

> **Further Comments on Shell Faces**
>
> When we employed shell elements in *Chapter 5, Analyses of Axisymmetric Bodies*, (in the sub-section *Assigning thickness*) we did not have to flip the face of the shell during the shell definition. You may be wondering why. Mainly because we created the surface from the inner face of the revolved body (you may refer to *Figure 5.12*). SOLIDWORKS will treat the face from which the thickness is created as the positive face, and hence the face to which loads, boundary conditions, and so on, should be applied. In the current case, what we have is just a plane surface with no pre-defined positive or negative faces, so we need to manually ensure the right face is referenced as the top or bottom.

It is now time to assign materials to the structure.

Creating and assigning a custom concrete material property

We shall start by assigning the material property to the floor slabs (made of concrete according to the problem statement). Regrettably, the SOLIDWORKS library does not contain a concrete material, which means we need to create it.

To this end, let's create and apply the concrete material properties as outlined next:

1. Within the simulation study tree, select all six surface bodies and right-click within the selected area of the selected items.

2. Click **Apply/Edit Material....**

    From the materials database that appears, take the next set of actions.

3. Scroll down to the material folder named `Custom Materials`, right-click, then click **New Category** (*Figure 7.8a*). In response to this, a new material folder will be created with the default name being `New Category`.

4. Change the name to `MyConcrete` (*Figure 7.8b*).

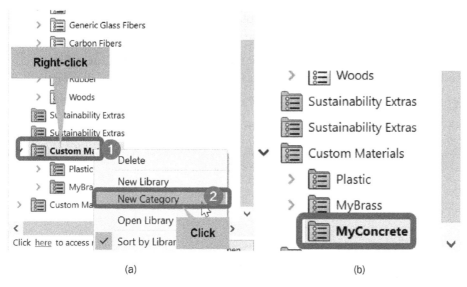

Figure 7.8 – Navigating to the custom material section

5. Next, right-click on the newly created `MyConcrete` folder and select **New Material** from the context menu that appears (*Figure 7.9a*). With this step, a new material file will be created with the default name `Default`. Change the name to `Concrete` as shown in *Figure 7.9b*.

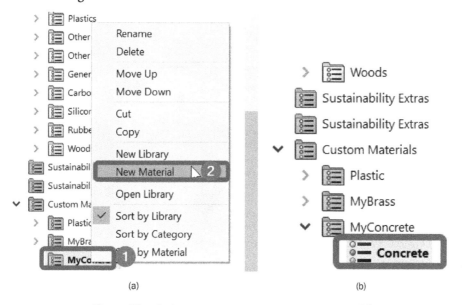

Figure 7.9 – Actions to create a custom concrete material

At this juncture, we have created the folder and the file for the concrete material. We are now ready to supply all the necessary values.

6. Within the **Material** database window, for **Model Type**, select **Linear Elastic Isotropic** and ensure the unit remains as **SI - N/m^2 (Pa)** (labeled 2 in *Figure 7.10*).

7. For the failure criterion, select **Mohr-Coulomb Stress** (labeled 3 in *Figure 7.10*).

8. For the material property values, follow the details in the box labeled 4 in *Figure 7.10*.

9. Click **Save**, **Apply**, and **Close** (in that order).

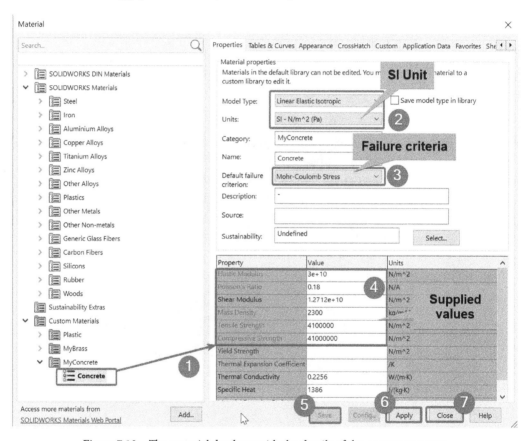

Figure 7.10 – The material database with the details of the custom concrete

It is important to point out we have selected the **Mohr-Coulomb Stress** failure theory in *step 7* to be consistent with a simplified status of concrete as a brittle material *[4]*.

> **A Note on Failure Criteria**
>
> You can also read about the various failure criteria in SOLIDWORKS Simulation on the help page: `https://help.solidworks. com/2022/english/SolidWorks/cworks/Failure_ Criteria.htm?verRedirect=1.`

With *steps 1–9* completed, we are done with the creation and application of the concrete material properties. As usual, a check mark should appear on the six surface bodies in the simulation study tree.

Next, we shift to the application of the material property to the beams and the columns.

Assigning material details to the beams/columns

Since the ASTM A36 material comes with SOLIDWORKS, the procedure here is straightforward. The steps to assign the ASTM A36 material property to the beams and columns are as outlined:

1. Right-click on the `Cut-List` folder and choose **Apply/Edit Material...** (*Figure 7.11*).

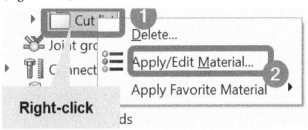

Figure 7.11 – Assigning a material to the bodies

2. From the materials database that appears, expand the `Steel` folder, then click on **ASTM A36 steel**.

3. Click **Apply** and **Close**.

The material assignment task is now completed. In the next sub-section, we will briefly take a look at the contact setting (before we deal with the application of the load and the specification of the fixture).

Examining the default contact setting

Due to the presence of multiple interacting parts, interaction conditions must be defined between them. By expanding the `Connections` folder as shown in *Figure 7.12*, you will notice that SOLIDWORKS has applied a global interaction (**Bonded**) at the assembly level for us.

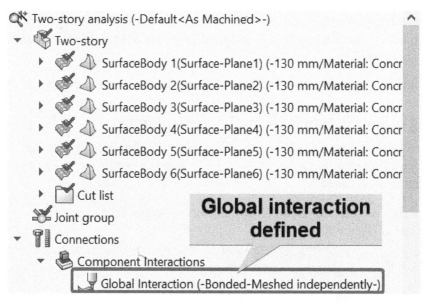

Figure 7.12 – Examining the interaction details

For this case study, the **Bonded** option is sufficient to hold all the parts together as a single piece, and there's no need to override it (later on in case study 2, we will need to go beyond the default contact). With this, we are done with the examination of the contact setting, and we can now transition to the application of the fixture.

Applying fixtures at the base of the lower columns

For the boundary conditions, otherwise known as fixtures within the SOLIDWORKS Simulation environment, we shall constrain the base of the building. To achieve this, follow the next set of steps:

1. Right-click on **Fixtures** and pick **Fixed Geometry** from the context menu that appears.

2. When the **Fixture** property manager appears, click on the Joints symbol (labeled 2 in *Figure 7.13*).

3. Navigate to the graphics window and click on the six bottom joints as shown in *Figure 7.13*.

4. Click **OK**.

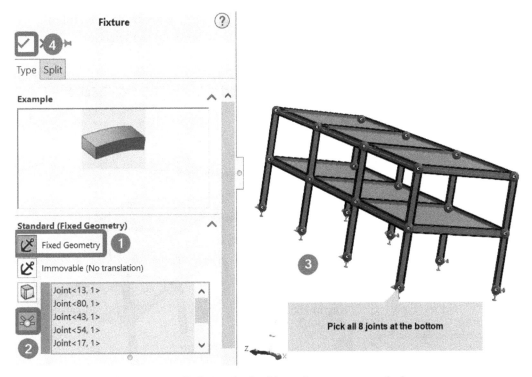

Figure 7.13 – Defining the fixed boundary supports at the base

With the completion of the application of the fixed boundary condition, let's now move on to the application of the loads.

Applying external loads

We have two loads to be applied to the building – the pressure and the gravitational loads. To apply these, follow the steps outlined next:

1. For the pressure load, right-click on **External Loads**, select **Pressure**, then follow the options indicated in *Figure 7.14*, and click **OK** to wrap up the selections.

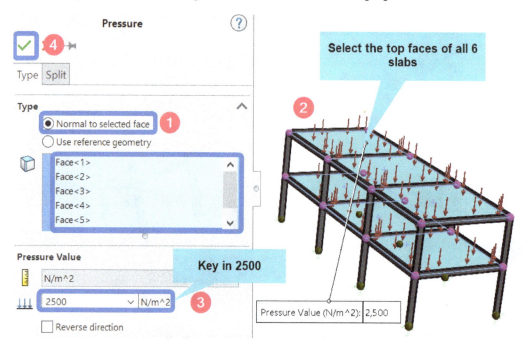

Figure 7.14 – Options for the application of the pressure load

2. For the gravitational load, right-click on **External Loads** and select **Gravity** (*Figure 7.15*).

Figure 7.15 – Initiating the application of the gravity load

Within the **Gravity** property manager that appears, execute the following actions.

3. Click inside the plane reference box (labeled 1 in *Figure 7.16*) and navigate to the graphics window to select **Top-Floor Plane**.

4. Check the **Reverse direction** box (labeled 3 in *Figure 7.16*), then click **OK** to wrap up the selections for the gravity load.

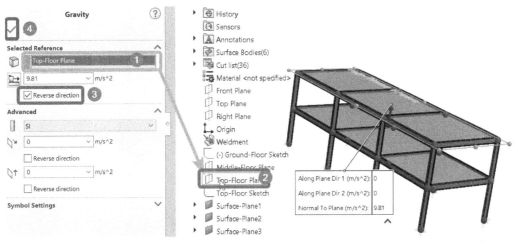

Figure 7.16 – Application of the gravity load

Over the past sub-sections, we have ticked a few items off the list of the preprocessing tasks. Specifically, we have seen how to define thickness for a surface created using a planar surface and how to flip its faces to have the right area for load application. We assigned material properties, which included steps to create a custom concrete material. Moreover, we've reviewed the contact settings, applied the fixture, and demonstrated the application of gravity and pressure loads. In the next sub-section, we shall transition to the meshing and post-processing phases of the workflow.

Part C: Meshing and running

In previous chapters, we have seen and discussed various ways to create a mesh. For this case study, the model involves mixed bodies (weldments for the beams and columns and surface bodies for the concrete slabs). Consequently, the meshing engine will discretize the beams/columns using beam elements, while the slabs will be discretized using shell elements as will be demonstrated in the sub-section that follows.

It is possible to create a mesh without giving thought to how it is done by simply using the **Create Mesh** command. But due to the fact that we have mixed elements for this case study, the approach we will adopt is to create mesh controls for the beam elements and use the default mesh setting for the shell elements.

To create the mesh controls and the mesh, take the following steps:

1. Right-click on **Mesh** and select **Apply Mesh Control**.

2. From the **Mesh Control** property manager that pops up, under **Selected Entities**, activate the beam option (labeled 1), then follow the other illustrations in *Figure 7.17*, and wrap up the selections by clicking **OK**. You are encouraged to use the box selection approach when selecting the beams.

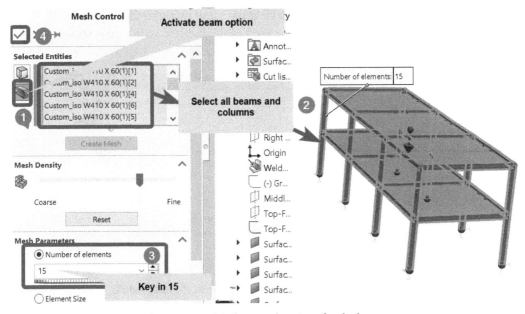

Figure 7.17 – Mesh control options for the beams

The box labeled 3 in *Figure 7.17* implies that we shall be discretizing all beams and columns with 15 elements each. Next, we generate the mesh for the whole assembly.

3. Right-click again on **Mesh**, then select **Create Mesh**. When the **Mesh Control** property manager appears, keep the options as shown in *Figure 7.18*, and wrap up the selections by clicking **OK**.

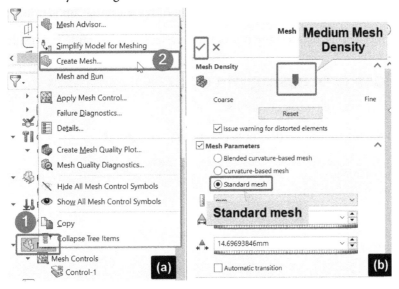

Figure 7.18 – General meshing options

4. After the running is complete, you can view the mesh by right-clicking on **Mesh** and then selecting **Show Mesh**. The top view of the meshed structure is shown in *Figure 7.19a*.

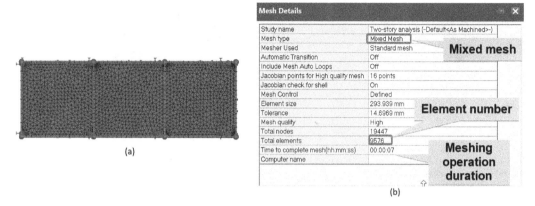

(a)

(b)

Figure 7.19 – (a) Top view of the meshed structure; (b) mesh details

5. Before we process the results, you can view the mesh details by right-clicking on **Mesh** and then selecting **Details**. The meshed detail is shown in *Figure 7.19b*.

6. To obtain the results, right-click on the analysis name and select **Run**. It will take some time to finish running.

Looking at the mesh details in *Figure 7.19b*, we can infer that: (i) the mesh quality is moderate; (ii) we have a mixed mesh with a total of 9,576 elements created via the *Standard Meshing* engine; and (iii) and the duration of the meshing run is minuscule.

With the activities about the meshing out of the way, let's now examine the results.

Part D: Post-processing of results

We aim to address the following questions:

- What is the maximum vertical deflection of the structure?

- What is the nature of the bending stresses in the beams/columns?

Let's kick off with the first question.

Obtaining the displacement result

To obtain the vertical displacement, we follow these steps:

1. Right-click on **Results**, then select **Define Displacement Plot**.

2. Within the **Displacement plot** property manager that pops up, under the **Definition** tab, navigate to **Display** and select **UY: Y Displacement** as highlighted in *Figure 7.20a*.

3. Also, move to the **Chart Options** tab and make the selections indicated in *Figure 7.20b*.

4. Click **OK** to finish off the selections.

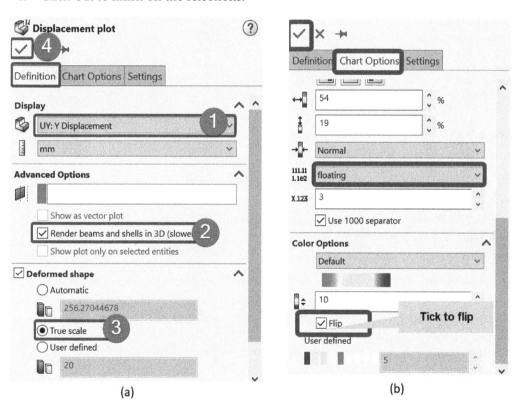

(a)   (b)

Figure 7.20 – Specifying options for the displacement plot

Figure 7.21 depicts the vertical deformation (which is the displacement along the Y-axis) of the whole structure.

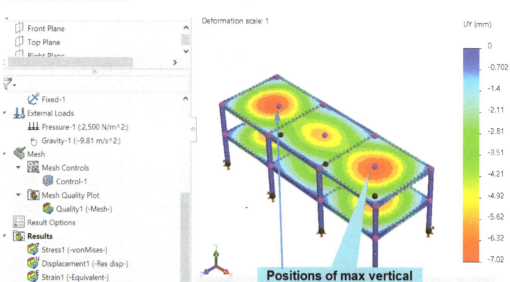

Figure 7.21 – The plot of the vertical deflection of the structure

It can be seen from *Figure 7.21* that a maximum vertical deflection of *-7.02 mm* is obtained as a consequence of the combined effect of a live load of 2,500 N/m² acting on the slabs and the self-weights of the whole structure. Besides that, the figure also signals that the location of the maximum deflection is at the center of the right and left upper slabs.

As much as we would like to move on to the next result, it is important to ask: How reliable is the value of displacement we have obtained? Before answering this question, recall that it has been pointed out in the past chapters that a simulation study should ideally be verified and validated before deriving technical conclusions from it. The validation could be through theoretical calculation, experiments, or data from other reliable sources.

Interestingly, unlike the past chapters, the complexity of the analyzed structure makes it difficult to have a closed-form solution via a theoretical framework to validate the value we have obtained. This kind of situation is prevalent in so many practical situations. And under such circumstances, an experiment may have to be conducted on a scaled-down version of the model to ascertain the accuracy of the simulation results. However, in the present case, in place of conducting an experiment, a similar analysis was carried out in Ansys (another simulation software). *Figure 7.22* depicts the displacement plot/result from Ansys.

C: Static Structural
Directional Deformation
Type: Directional Deformation(Y Axis)
Unit: mm
Global Coordinate System
Time: 1
Max: 0.0021
Min: -7.195

0.0021
-0.7975
-1.597
-2.397
-3.196
-3.996
-4.796
-5.595
-6.395
-7.195

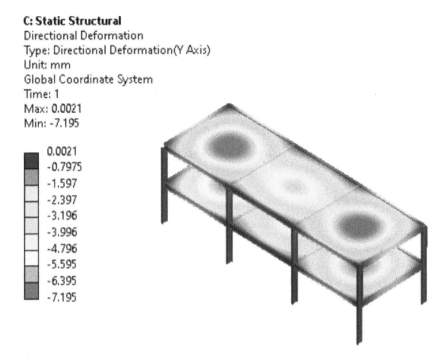

Figure 7.22 – Vertical deflection of the structure from Ansys

As you can see, the maximum displacement results from SOLIDWORKS Simulation (in *Figure 7.21*) and the results from Ansys (in *Figure 7.22*) only differ by around 2.4% despite using a moderately defined mesh. Just to be clear, it is not a normal practice to have to do this comparison with other software (especially in companies with limited resources). We have done this comparison just to reinforce our confidence in the reliability of the SOLIDWORKS Simulation result.

Meanwhile, we have focused on the vertical deflection here because the two loads are directed along the Y-axis. In a more elaborate analysis, it may be desirable to include the effect of wind load in a specific direction, say along the Z-axis. For such a scenario, it is best to look at the resultant displacement of the whole building. Besides, if a seismic analysis is the purpose of the simulation, then a quasi-static or dynamic analysis may have to be conducted.

Relatedly, we may also uncover further details about the structure from the stresses that develop within the members. For this reason, we will now proceed to examine the stress results in the next sub-section.

Extracting the stresses in the components

We shall restrict ourselves to the bending stresses in the beams/columns. In this vein, follow the steps given next to obtain the bending stresses:

1. Right-click on the `Results` folder, then select **Define Stress plot**.

2. Within the **Stress plot** property manager that appears, under **Display**, activate the **Beams** option, then follow the selections shown in *Figure 7.23*.

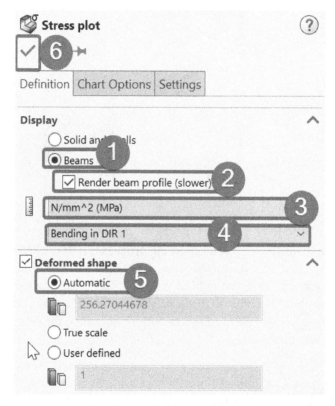

Figure 7.23 – Specifying options for the bending stress along direction 1

3. Now, move to the **Chart Options** tab. Under **Position/Format**, change the number format to **floating**.

4. Click **OK**.

With *steps 1–4* completed, the graphics window is updated with the plot of the bending stress along direction 1 (X-axis) as depicted in *Figure 7.24*. As you will notice from the stress plot, the compressive and tensile bending stresses in the members are roughly *-38 MPa/+38 MPa*. These values are far below the yield strength of ASTM 36 steel, from which the members are made, and will consequently yield acceptable von Mises stress.

Plot type: Bending in DIR 1 Stress2
Deformation scale: 256.27

Bending in DIR 1 (N/mm^2 (MPa))

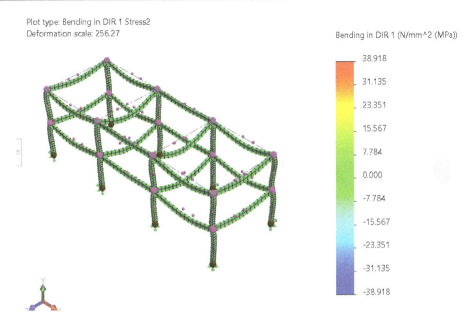

38.918

31.135

23.351

15.567

7.784

0.000

-7.784

-15.567

-23.351

-31.135

-38.918

Figure 7.24 – The deformed shape of the beams for the bending stress along the X-axis

Moreover, *Figure 7.24* signifies that the exterior columns are the most affected by the bending stress along the X-axis.

Next, if you repeat *steps 1–4* to obtain the stress along direction 2 by choosing **Bending in DIR 2** (see *Figure 7.23*), the result should be similar to *Figure 7.25*.

Deformation scale: 256.27

Bending in DIR 2 (N/mm^2 (MPa))

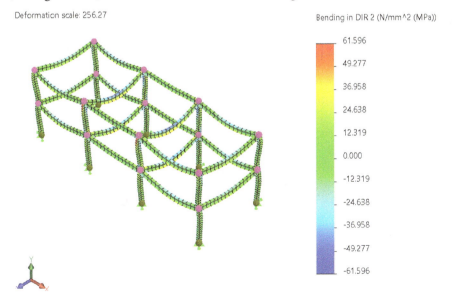

61.596

49.277

36.958

24.638

12.319

0.000

-12.319

-24.638

-36.958

-49.277

-61.596

Figure 7.25 – The deformed shape of the beams for the bending stress along the Z-axis

Figure 7.25 reveals that when the stress is examined along direction 2 (the Z-axis), the interior beams of the upper floor are the most stressed. Besides, the maximum compressive and tensile bending stresses in the members are roughly *-61 MPa/+61 MPa*. Likewise, these values are not too high to cause concern. Now, although not covered, it is recommended that you scrutinize the stresses in the slabs as well. To sum up, note that the deformed configurations that are shown in *Figure 7.24* and *Figure 7.25* are just for illustrative purposes; they do not represent the true scale of the deformed configurations of the structure. All in all, the results we have discussed show that the response of the building is within acceptable design criteria. Now, assuming the structural deformation of the building is unsatisfactory, can you suggest ways to improve the design?

We have now completed the simulation and obtained the answers to the questions posed for the case study. In all, the procedure for the solution to the hands-on exploration has been a mix of things we have seen in the previous chapters, such as how to deal with material property assignment and examination of contact settings, and so on. But we also managed to demonstrate a few new things such as (i) the creation of a custom concrete material from scratch; (ii) the initiation of mesh control for each family of mesh types in a mixed element simulation task; and (iii) the combination of a gravitational load with the more traditional load type.

In the next case study, we will extend this problem further. In doing so, you will see how to stabilize a structure using SOLIDWORKS soft springs and how to create a no-penetration interaction between a family of **solid elements** and shell elements, among other things.

Analysis of three-dimensional components with mixed shell and solid elements

Here is the backstory for our second case study. As part of the drive toward decarbonization, the development of solar roof collectors has emerged as one of the simple measures toward the management of the energy crisis confronting our world. This is evident in some of the recent research studies, see for instance [5, 6]. In many cases, these roof-based solar modules are external add-ons to the buildings. This means the effect they have on the structural response of the building may not have been considered at the initial design stage. This case study, therefore, springs from an attempt to examine this effect.

Problem statement – case study 2

In the next few pages, we wish to extend the previous case study to account for the presence of three *3 m by 3 m* solar collectors on the top slabs as illustrated in *Figure 7.26*. Primarily, our objective is to evaluate the ensuing change in deformation and stress on the beams/columns due to the addition of the solar modules to the roof if they collectively impose a pressure load of 600 N/m².

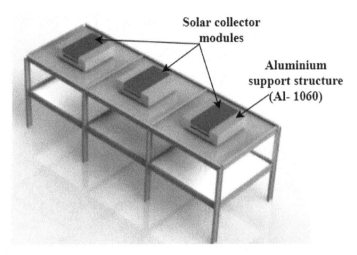

Figure 7.26 – The model of the two-story building with three solar collectors

Take note that in what follows, we will not include the nitty-gritty detail of the solar collectors. As a form of simplification, we will only consider the aluminum support structures upon which the pressure load from the solar collectors will be applied. Furthermore, we shall build on the foundations we have established so far to present a condensed exploration of the solution.

In all, this case study assumes that you have completed case study 1. This means the following:

- You have downloaded the Chapter 7 folder from the book website to retrieve the models.

- You have copied the Custom folder into the SOLIDWORKS parent directory for weldment profiles as described in the sub-section *Copying and reviewing the file in the custom profile* (for case study 1).

- You created the custom concrete materials as explained in the sub-section *Creating and assigning a custom concrete material property*.

If the preceding tasks have been completed, then we can begin with a review of the new model in the next sub-section.

Reviewing the model and activating the simulation study

Follow these steps to bring up the new model of the building with the integrated solar modules' support:

1. Start up SOLIDWORKS.

2. Select **File → Open**, then open the part file named `Two-story-Solar` (from the `Chapter 7` folder you downloaded).

3. Navigate to the feature manager tree and examine the details as shown in *Figure 7.27*.

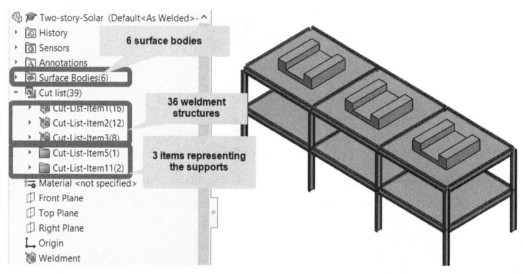

Figure 7.27 – Components of the building with the solar modules' support structures

Figure 7.27 shows that we now have an additional 39 items in the `Cut-List` folders. This is the sum of 36 structural beams/columns and the 3 new solar supports.

Apart from this, by hiding the three solar module supports, you will observe (as shown in *Figure 7.28*) that we have introduced the bearing area on the concrete slabs via the **Split line** tool. The essence of this is to allow us to apply the right component-level contact condition between the solar modules' support structures and the concrete slabs later on during the analysis. Recall that the **Split line** tool was introduced in *Chapter 6, Analyses of Components with Solid Elements*.

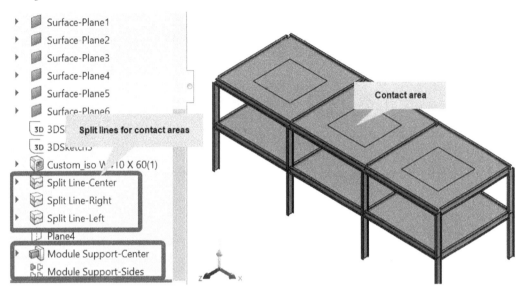

Figure 7.28 – Illustration of contact areas created with split lines

With the basics of the model established, let's now venture into the analysis environment by activating the simulation study as follows:

1. Click on **SOLIDWORKS Add-Ins**, then click on **SOLIDWORKS Simulation** to activate the **Simulation** tab.

2. With the **Simulation** tab active, create a new study with the name Two-story-Solar analysis (for a guide, see the sub-section *Activating the Simulation tab and creating a new study*).

Once the simulation study is launched, the simulation study tree should appear as shown in *Figure 7.29*.

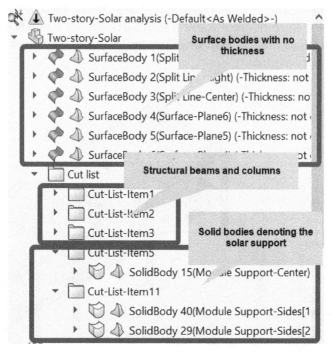

Figure 7.29 – Simulation property manager (building with solar module supports)

It might have caught your attention in *Figure 7.29* that when we mesh the structures, we will have shell elements (from the surface bodies), solid elements (from the solar supports), and beam elements (from the structural columns/beams). Additionally, as was the situation in case study 1, the surface bodies have no thickness yet. So, let's address the definition of the thickness and the material properties.

Defining surface thickness and assigning material properties

To assign thickness, the steps are akin to what we described earlier. For this reason, follow the procedure outlined in the sub-section *Assigning thickness* for case study 1 to assign thickness to the 6 surface bodies.

As for the specification of material properties, we have the surface bodies (made of concrete), the beams/columns (made of ASTM 36 steel), and solar module supports (derived from Aluminium alloy 1060). The steps to specify each of these are highlighted next.

For the surface bodies, do the following:

1. Select all six surface bodies and right-click within the selected area of the selected items as shown in *Figure 7.30*, then click **Apply/Edit Material....**

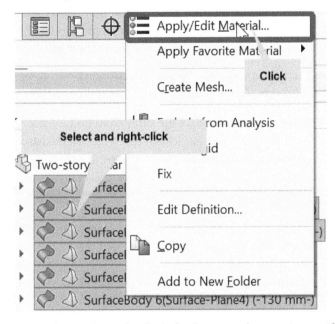

Figure 7.30 – Selecting the surface body for the material properties specification

2. From the material database that appears, navigate to the Custom Materials folder, click on Concrete, click **Apply**, then click on **Close** as presented in *Figure 7.31*.

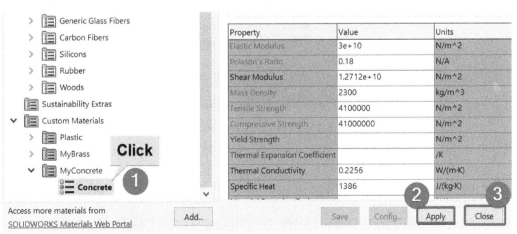

Figure 7.31 – Assigning concrete to surface bodies

With this, the concrete material properties are assigned to the surface bodies. Next, we resolve the remaining two materials as follows.

For the beams and columns, do the following:

Specify ASTM 36 steel by following the steps described in the sub-section *Assigning material details to the beams/columns* for case study 1. Note that under the current case, you need to apply material properties one by one on each of the three `Cut-List` folders for the beams/columns.

For the solar modules' supports, do the following:

1. Within the simulation study tree, select the three solid bodies that represent solar supports as shown in *Figure 7.32*, right-click within the selected area of the selected items, then click **Apply/Edit Material....**

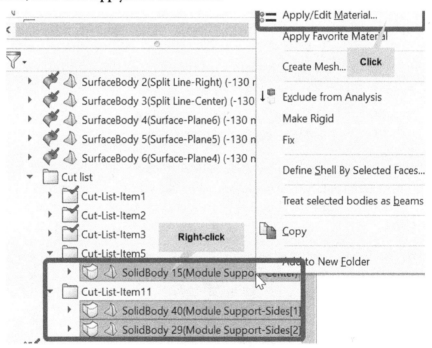

Figure 7.32 – Selecting the surface body for the material properties specification

2. From the materials database that appears, expand the `Aluminium Alloys` folder, then click **1060 Alloy**.

3. Click **Apply** and **Close**.

This wraps up the specification of the three material properties. Our next focus centers around contact settings.

Updating the default contact/interaction setting

The choice of the interaction setting will either make or break the current case study. Why? You will recall that for case study 1, we revealed that SOLIDWORKS has applied a global interaction setting – **Global Interaction (Bonded-Meshed independently-)** – at the assembly level for us (see the sub-section *Examining the default contact setting*). However, for the current case study, this global interaction is not enough because the solar module support structures are not connected to the concrete surface in the sense of "bonding." As a result, we will need to take things a bit further by defining a component-level interaction between the solar modules' support structures and the faces of the concrete slabs.

To enforce the new interaction settings, follow the steps highlighted next (carefully):

1. Within the simulation study tree, expand the `Connections` folder, right-click on **Component Interactions**, then select **Local Interaction** as depicted in *Figure 7.33*.

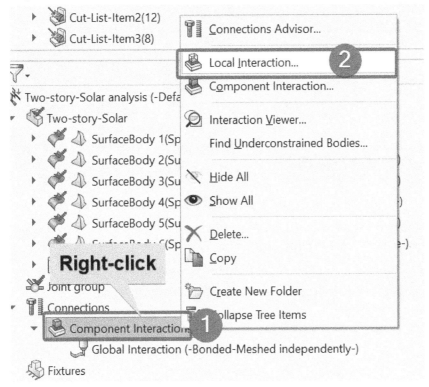

Figure 7.33 – Activating the local interactions

Within the **Local Interactions...** property manager that appears, we need to execute the next actions.

2.   Under **Interaction**, select **Automatically find local interactions**.

3.   Under **Options**, select **Find shell edge - solid/shell face pairs**, then navigate to the graphics window to select the three supports and the top surfaces one by one as shown in *Figure 7.34*.

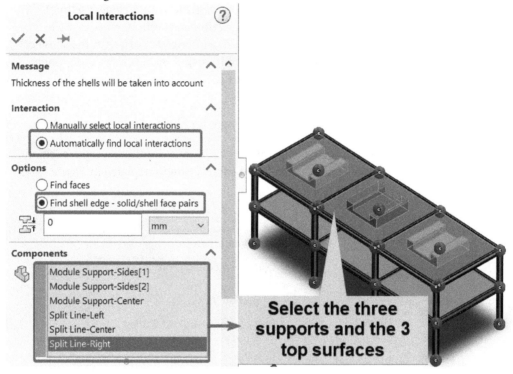

Figure 7.34 – Selecting the support structures for the local interactions

4.   Still under **Options**, with the six items (three solid bodies and three surfaces) selected, change back to **Find faces** *(Figure 7.35)*.

5.   Key in a value of 2 mm in the **Maximum Clearance** box as portrayed in *Figure 7.35*.

6.   Now click the **Find local interactions** button (see *Figure 7.35*).

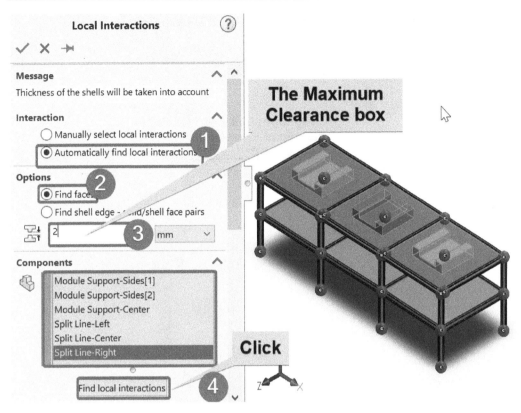

Figure 7.35 – Detecting the local interactions condition

The three local interactions between the supports and the slabs will appear in the **Results** box (labeled 1 in *Figure 7.36*). Scroll down to complete the rest of the actions.

7. Using the *Ctrl* key, select all three interactions within the **Results** box (labeled 1 in *Figure 7.36*).

8. Below the **Results** box, for **Type**, ensure the option is set to **Contact** (labeled 2 in *Figure 7.36*).

Figure 7.36 – Finalizing the local interactions condition

9. Under **Advanced**, ensure the option is set to **Surface to surface** (labeled 3).

10. Click on the green add button beside the **Results** box (labeled 4).

11. Click **OK**.

12. After clicking **OK**, you may get a warning that says **Do you want to create the selected local interactions?**. Click **Yes**.

With *steps 1–12* completed, the Connections folder will update to show the new interactions as manifested in *Figure 7.37*.

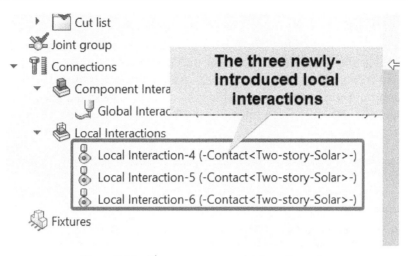

Figure 7.37 – The new component interaction sets

Although we briefly highlighted the importance of interactions in *Chapter 6, Analyses of Components with Solid Elements*, we did not give much detail about them. So, let's step back a bit to look at the interaction definitions for this case study. First, as you can see from *Figure 7.37*, we now have two different interaction conditions under the Connections folder. Let's highlight the difference between the two:

- *Component Interactions*: Technically, a component interaction setting defines the contact condition between selected components. Under the component interaction property manager, you may choose to define a global interaction setting for all components in a multi-body component/assembly or a narrower interaction condition between specific components. Indeed, we have adopted **Global Interactions (Bonded)** in the current study. This holds all the bodies or components together as one piece and it is the default interaction/contact condition for multi-body parts. In situations where this default interaction condition may not be enough, you must define local interactions between selected components.

- *Local Interactions*: As you can see, in this case study, we have manually defined **Local Interactions** between specific geometric entities, specifically the faces of bodies that are interacting. It is pertinent to know that a local interaction can be specifically chosen to be any of these: **Contact, Bonded, Shrink fit, Free,** and **Virtual Wall**. Obviously, we have used the **Contact** option here. The **Contact** interaction generally averts interference between two sets of entities. This means the entities cannot penetrate each other if an external force brings them together, but they can move away from each other. You can explore more information about these options at the following SOLIDWORKS help link: https://help.solidworks.com/2022/english/SolidWorks/cworks/IDH_HELP_FIND_CONTACT_SETS.htm?verRedirect=1.

One more thing before we move on. You will notice that in *Figure 7.35*, a value of 2 mm was specified inside the **Maximum Clearance** box. This value can be changed depending on the estimated gap between the surfaces. For the current scenario, a gap of 0 . 2 mm was maintained between the bearing area of each concrete slab and the face of the solar module's support during modeling. Since this gap falls below the stated maximum clearance value, SOLIDWORKS can automatically determine the contact pairs for the interacting faces.

Important Note

If you are using an earlier version of the SOLIDWORKS Simulation, take note of the following changes in terminology:

Component Interactions in the 2021-2022 version of SOLIDWORKS used to be called **Component Contacts** in the earlier versions of SOLIDWORKS. Also, **Local Interactions** in the 2021-2022 version of SOLIDWORKS used to be referred to as **Contact Sets** in the earlier versions. Moreover, under the **Contact Sets** property manager (in the earlier versions of SOLIDWORKS), the type of contacts you will see are **Bonded**, **Shrink fit**, **Allow Penetration**, **No Penetration**, and **Virtual Wall**. In contrast, in the 2021-2022 version, these options are now known as **Contact**, **Bonded**, **Shrink fit**, **Free**, and **Virtual Wall**, respectively.

We are done with updating the interaction, and we can now switch to the application of the fixtures and the loads.

Applying fixtures and loads

The fixture to be applied is the same as what we specified for case study 1. That is, to fix all the joints at the base of the building. For a guide on this, see the sub-section *Applying fixtures at the base of the lower columns* in case study 1.

Once the application of the fixture is complete, we move on to the specifications of the external loads.

Mainly, we have three sets of loads to apply (gravity on the whole assembly, pressure load on the surface bodies, and pressure load on the solar module support structures).

For the gravity load, follow the procedure used to apply it to the structure as described for case study 1 in the sub-section *Applying the external loads*.

To implement the pressure load on the concrete slabs, here are the steps:

1. First, navigate to the feature manager tree, then hide each of the support structures one by one. *Figure 7.38a* illustrates the action for the middle support structure.

2. With the support structures hidden, move back to the simulation study tree, right-click on **External Loads**, select **Pressure**, then follow the options indicated in *Figure 7.38b*, and click **OK** to wrap up the selections.

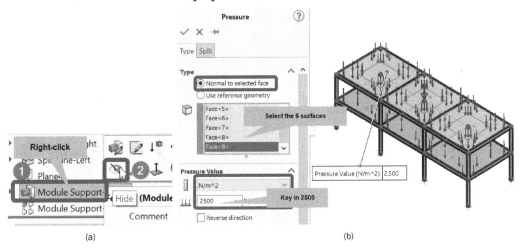

(a)   (b)

Figure 7.38 – Preparation for the surface pressure application

The last load to be applied is the one acting on the support structures. For this, follow the next steps to apply the pressure loads:

1. First, ensure the support structures are unhidden in the feature manager design tree.

2. Next, right-click on **External Loads**, select **Pressure**, then follow the options indicated in *Figure 7.39*.

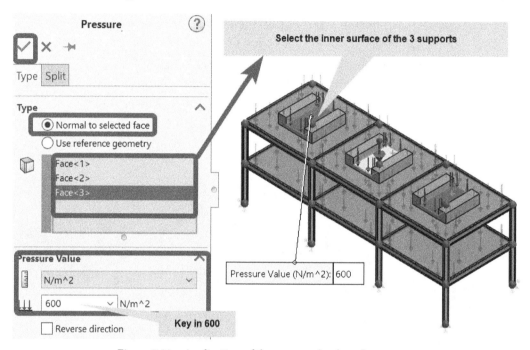

Figure 7.39 – Application of the pressure load on the support

Put together, we have now completed the assignment of material properties to all bodies and ensured the fixtures and the loads are applied. We have also assigned thickness to the surface bodies and imposed an interaction/contact condition between selected entities (in addition to the global bonded interaction definition). We are now set to pursue the crucial meshing operations in the next section.

Meshing

The partial view of the simulation study tree shown in *Figure 7.40* implies that we have a combination of surface bodies, solid bodies, and beams. As a result, the discretization will comprise shell elements, solid elements, and beam elements.

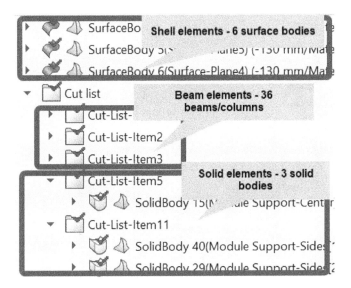

Figure 7.40 – A combination of surface, beam, and solid bodies

Similar to our approach for case study 1, we shall first create mesh control for each family of elements. Afterward, we will generate the mesh. The steps to create the mesh controls for each family of bodies are outlined as follows.

Here are the steps for the mesh control for the beam elements:

1. Right-click on **Mesh**, then select **Apply Mesh Control**.

2. From the **Mesh Control** property manager that appears, under **Selected Entities**, activate the beam option, then follow the instructions in *Figure 7.41*.

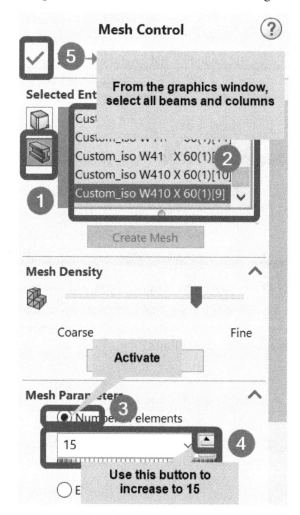

Figure 7.41 – Mesh Control options for the beams

Notice that we are again instructing the meshing engine to discretize each beam/column with 15 beam elements. The simulation study tree should now appear as depicted in *Figure 7.42*:

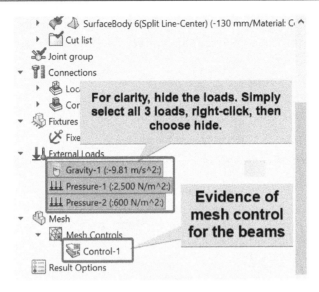

Figure 7.42 – Updated view of the simulation study tree

Here are the steps for the mesh control for the solar supports:

1. If hidden, then unhide the solar modules' support from the feature manager design tree.

2. Next, implement the **Mesh Control** as illustrated in *Figure 7.43*.

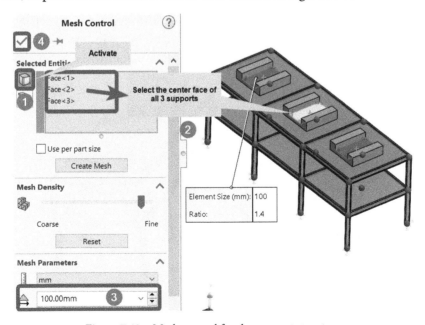

Figure 7.43 – Mesh control for the support structures

With the mesh control specifications accomplished, you should check the simulation study tree to ensure the presence of the controls as indicated in *Figure 7.44*.

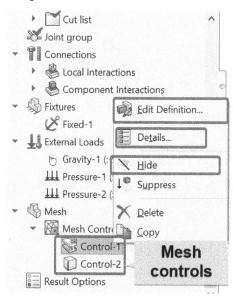

Figure 7.44 – The two mesh controls

Note that each of the mesh control definitions can be examined by simply right-clicking, through which you can also choose to work with the **Edit Definition**, **Details**, **Hide/Show** options, and so on. The **Hide/Show** option is especially useful to prevent overcrowding the model details in the graphics window. For instance, *Figure 7.45* reflects the difference between showing and hiding the mesh controls.

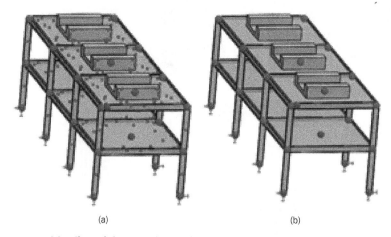

(a)   (b)

Figure 7.45 – (a) Effect of showing the mesh controls; (b) Effect of hiding the mesh controls

Note that a similar option (that is, **Hide/Show**) is available for the external loads and fixtures as well. Indeed, the symbols for all three loads that we have applied early on have been hidden to facilitate a neat graphics window (as indicated in *Figure 7.42*).

With the three meshing controls defined, let's now create the mesh by following the steps highlighted next:

1. Right-click on **Mesh**, then select **Create Mesh**.

2. From the **Mesh** property manager that pops up, increase the fineness of the mesh using the Mesh Factor button as shown in *Figure 7.46a*.

3. Accept the other default options and click **OK**.

    After the meshing is complete, the discretized structure should appear as shown in *Figure 7.46b*.

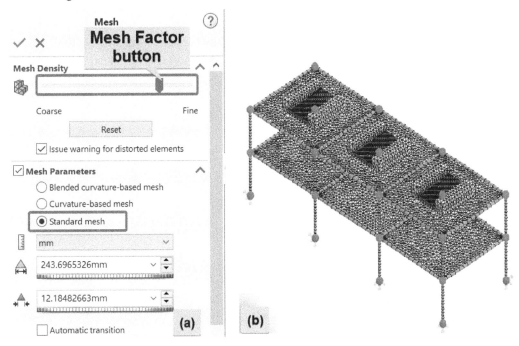

Figure 7.46 – (a) The mesh property manager; (b) View of the meshed structure

We are now poised to run the analysis and retrieve the desired results.

Running the analysis and post-process

Throughout the previous chapters (including the first case study), we have always taken the running of the analysis to be straightforward by accepting the default running properties. But for the current case study, we need to handle things a bit differently for the following reasons:

- First, we have contact interaction (previously known as no penetration contact) between the solar supports and the concrete slabs. A contact interaction is essentially a non-linear interaction condition that will often elongate the duration of the simulation.

- Second, given that the solar modules' support structures are not considered bonded to the surface of the slabs, they may slide or experience a rigid body motion when loads act upon them. Consequently, to prevent this rigid body motion, we shall use one of the SOLIDWORKS Simulation features called *soft springs*.

To address the issues raised above, we shall now carry out the following steps:

1. Right-click on the analysis/study name, then select **Properties...** as shown in *Figure 7.47*.

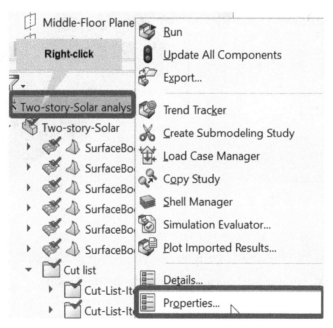

Figure 7.47 – Activating the study property manager

2. From the static study property manager that appears, under **Solver**, ensure the **Selection** option is set to **Automatic** (labeled 1 in *Figure 7.48*).

3. Next, check the **Use soft spring to stabilize model** option (labeled 2 in *Figure 7.48*).

4. Click **OK** to close the static study property manager.

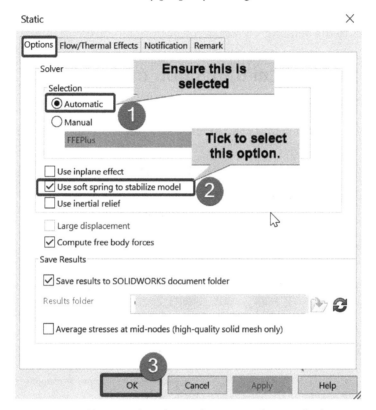

Figure 7.48 – Changing the solver and activating the use of soft spring

5. Finally, after the manager window is closed, right-click again on the study name, then select **Run** as depicted in *Figure 7.49*.

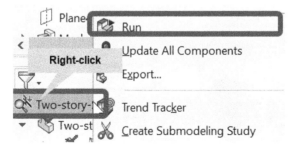

Figure 7.49 – Running the analysis

By monitoring the status of the running process, you will see that it takes a significantly longer time to complete the process than any of the case studies we have studied so far. In any case, after the completion of the running process, we are now set to see what the results tell us about the response of the structure. Our major interests are as follows:

- The new value of the maximum vertical deflection of the concrete slabs

- The changes to the bending stresses of the beams/columns

Obtaining the vertical deflection of the building

Obtain the vertical displacement of the building by following these steps:

1. Navigate to the feature manager to hide the solar modules' support structures (similar to *Figure 7.38a*).

2. Next, right-click on **Results**, then select **Define Displacement Plot**.

3. From the **Displacement plot** property manager that appears, under the **Definition** tab, navigate to **Display** and select **UY: Y Displacement**.

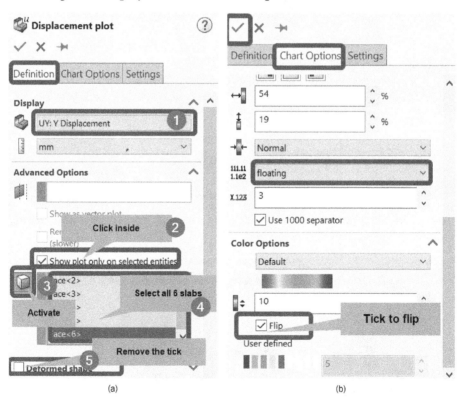

(a)   (b)

Figure 7.50 – Options to retrieve the slabs' vertical displacement

4. Update the other usual options by referring to the pictorial guide in *Figure 7.50b*.

5. Click **OK** to finish off the selections.

By completing *steps 1–4*, the plot of the distribution and pattern of the deformation of the slabs along the direction of the Y-axis is revealed as depicted in *Figure 7.51*.

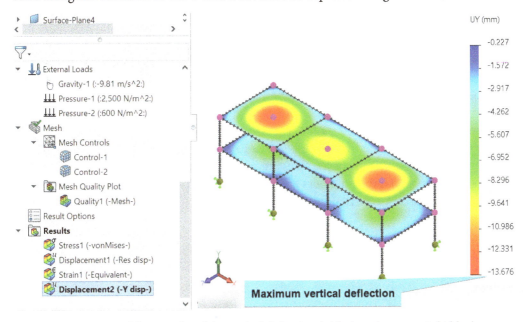

Figure 7.51 – The new plot of the vertical deflection (with the solar supports hidden)

Figure 7.51 suggests that the addition of the three 3 m by 3 m solar collectors leads to a maximum vertical deflection of *-13.676 mm*. This represents more than a 70% increase in the vertical deformation of the slab compared to case study 1. While the value of 600 N/m² pressure imposed on the solar supports represents an approximate hypothetical value, this case study highlights the significant effect that external add-ons acting as a dead load can have on the response of such a structure.

Oftentimes, the serviceability criteria for beams supporting floors have to be checked as well. Hence, it is recommended that you explore the deformations of the beams and columns. In the next sub-section, we shall take a look at the bending stresses that develop in the beams and compare the values to what we obtained in the first case study.

Extracting the stresses in the components

Retrieve the bending stress along direction 1 by following the steps described next:

1. Right-click on the `Results` folder, then select **Define Stress plot**.

2. Within the **Stress plot** property manager that appears, under **Display**, activate the **Beams** option (if need be, see the pictorial guide in *Figure 7.23*) and select **Bending in DIR 1**.

3. Now, move to the **Chart Options** tab. Under **Position/Format**, change the number format to **floating**.

4. Click **OK**.

In response to the completion of *steps 1–4*, the stress plot is displayed as depicted in *Figure 7.52*, from where it is seen that the bending stresses along direction 1 (X-axis) are now about *-59 MPa/+59 MPa*. While these values are still well below the yield strength of ASTM 36 steel, from which the members are made, they represent about a 50% increase in the stress along direction 1 reported for case study 1.

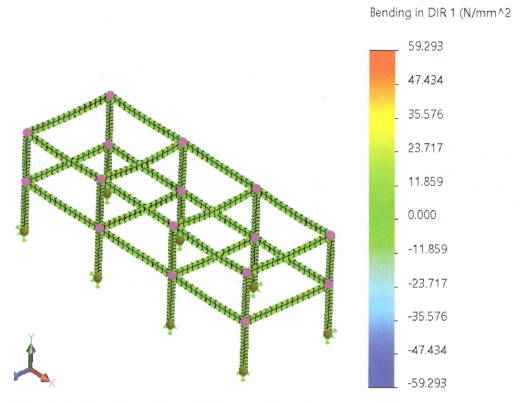

Figure 7.52 – The deformed shape of the beams for the bending stress along the X-axis (case study 2)

Repeat *steps 1–4* to examine the stress along direction 2 by choosing **Bending in DIR 2**. The result is presented in *Figure 7.53*.

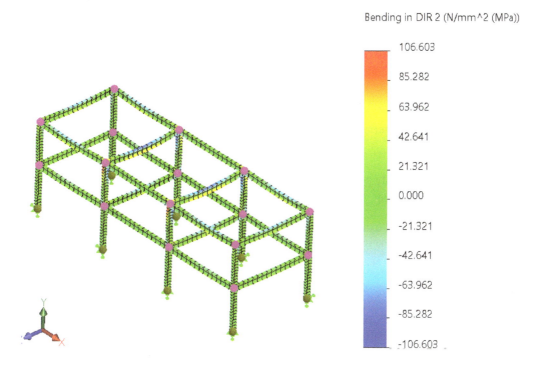

Bending in DIR 2 (N/mm^2 (MPa))

106.603
85.282
63.962
42.641
21.321
0.000
-21.321
-42.641
-63.962
-85.282
-106.603

Figure 7.53 – The deformed shape of the beams for the bending stress along the Z-axis (case study 2)

As you can see in *Figure 7.53*, the maximum compressive and tensile bending stresses in the members are now *-106 MPa/+106 MPa*. This is also about 50% higher than was reported for case study 1.

This ends our exploration of the second case study and marks the end of this chapter. In tackling the two simulation studies, we have harnessed a few more features of SOLIDWORKS Simulation. Like the other chapters, there are far more results that we could obtain from the simulation studies. It is hoped that you have gained a foundation to go deeper and check some of these other results. For instance, the factor of safety of the beams/columns, the factor of safety of the slabs, the strains and stresses in the slabs, and so on.

Summary

We have explored two examples within this chapter that allowed us to walk through simulations with mixed elements. Via these case studies, we have demonstrated the following:

- How to handle local mesh control for each family of elements in a mixed element analysis

- How to create a custom material from scratch and apply gravity load

- How to define a no penetration contact set between solid and shell bodies

- How to use the automatic contact pair detection tool

- How to update a study property to select a solver and employ the in-built soft spring to stabilize a structure

In the next chapter, we will examine the analysis of components with advanced materials in the form of composites.

Exercises

1. Repeat case study 2 by replacing the *contact local interaction* with a *bonded local interaction*, in addition to the global bonded interaction. What effect does this have on the running of the simulation, the deformation, and the stress distributions?

2. Repeat case study 2 for a situation where the thickness of the concrete slabs at the top is changed to 160 mm, and those at the bottom changed to 100 mm. Re-conduct the simulation without the use of the *soft spring* option. Explain how these changes affect the simulation running behavior, the deformation, and stress distributions.

Further reading

- [1] *Simulation - Mixed Mesh Techniques, S. Solutions, ed, 2019, pp.* `https://www.solidsolutions.ie/solidworks-videos/simulation-mixed-mesh-techniques.aspx.`

- [2] *H. H. Lee, Finite Element Simulations with ANSYS Workbench 17, SDC Publications, 2017.*

- [3] *Design Loads for Residential Buildings, US, 2000 [Online]. Available:* `https://www.huduser.gov/publications/pdf/res2000_2.pdf`

- [4] *An Introduction to Soil Mechanics, A. Verruijt, Springer International Publishing, 2017.*

- [5] *Design, construction and performance prediction of integrated solar roof collectors using finite element analysis, M. M. Hassan and Y. Beliveau, Construction and Building Materials, vol. 21, no. 5, pp. 1069-1078, 2007.*

- [6] *Performance evaluation and techno-economic analysis of a novel building integrated PV/T roof collector: An experimental validation, B. Mempouo, and S. B. Riffat, M. S. Buker, Energy and Buildings, Vol. 76, pp. 164-175, 2014.*

Section 3: Advanced SOLIDWORKS Simulation with Complex Material and Loading Behavior

This section is dedicated to some advanced features of SOLIDWORKS Simulation. Among other things, this section provides you with knowledge of fatigue failure of machine components, analysis of advanced components made of composite materials, investigation of the thermo-structural performance of components, and the awareness of the optimization capability of SOLIDWORKS Simulation for reliable designs of engineering components against failure.

This section comprises the following chapters:

- *Chapter 8, Simulation of Components with Composite Materials*
- *Chapter 9, Simulation of Components under Thermo-Mechanical and Cyclic Loads*
- *Chapter 10, A guide to Meshing in SOLIDWORKS*

8
Simulation of Components with Composite Materials

So far in this book, the simulation examples presented have involved components constructed of isotropic materials – that is, materials for which the properties such as Young's and shear moduli and Poisson's ratio are constant in all directions. However, given that there has been an increase in the use of composite materials in many industries in recent years, knowledge of simulation with these materials has become central to the design of advanced engineering products. For this reason, this chapter taps into SOLIDWORKS's simulation capability for the structural analysis of components made of composite materials. The chapter covers the following topics:

- An overview of the analysis of composite structures
- An analysis of composite beams
- An analysis of advanced composite structures

Technical requirements

You will need to have access to the SOLIDWORKS software with a SOLIDWORKS Simulation license.

You can find the sample file of the model required for this chapter here:

```
https://github.com/PacktPublishing/Practical-Finite-Element-
Simulations-with-SOLIDWORKS-2022/tree/main/Chapter08
```

An overview of the analysis of composite structures

Before we go into the case study for this chapter, it helps to have some background knowledge about composite materials. Engineering materials such as metals, ceramics, and polymers/elastomers are isotropic by nature. Although each of these families of materials is widely used in their natural isotropic form for many applications, each has its pros and cons. For instance, solid metals (as opposed to liquid metals such as gallium and mercury) are endowed with high stiffness and have high deformability, but are heavy and prone to fatigue failure. Ceramics also have high stiffness, but they are brittle. On the other hand, polymers/elastomers have good corrosion and wear-resistance properties, but they have very low stiffness and temperature-dependent properties. Put together, this indicates that no single class of material is superior under all possible desirable functional working conditions *[1]* (see the *Further reading* section at the end of the chapter). In response to the weaknesses of each material family as outlined, research into composite materials has become a key driver of many engineering innovations in recent years.

In the most basic form, a composite is formed when two or more different materials are combined. This combination may involve microscopic or macroscopic combinations. Further, the combination may occur naturally (for instance, wood and bone natural composites) or be artificially engineered. Our interest in this chapter is those that involve artificially engineered macroscopic material combinations. The applications of this class of artificially engineered composite materials can be found in space/aircraft structures, automotive systems, bicycle frames, golf club shafts, energy systems such as wind turbines, and so on. *Figure 8.1* illustrates some practical and modern usage of composite materials:

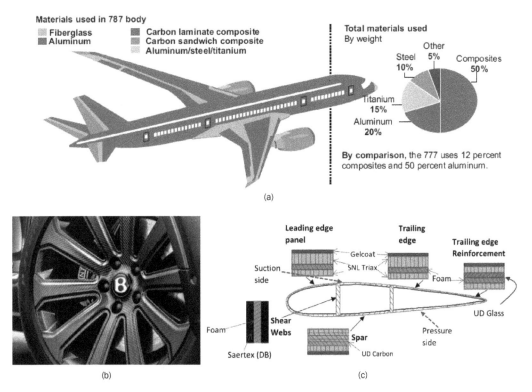

Figure 8.1 – An illustration of the usage of composites [2, 3]: (a) for aircraft bodies, (b) a carbon fiber-reinforced composite wheel, and (c) a cross section of a composite wind turbine blade

Notably, there are also various kinds of artificially engineered composite materials, as depicted in *Figure 8.2*. A detailed discussion of the difference in the constituent, behavior, properties, and manufacturing of each category in *Figure 8.2* can be found in the book by *Hyer and White [4]*:

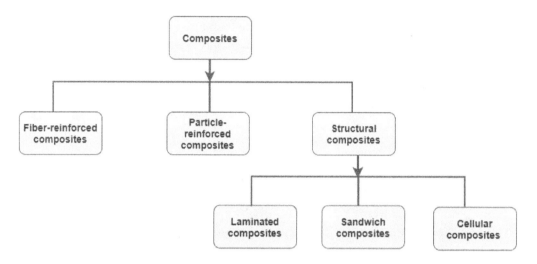

Figure 8.2 – Some major categories of composites

As you will see in sections ahead, much of this chapter centers on the analysis of components designed with laminated composites formed from fiber-reinforced composites. In laminated composites, the laminate is formed by stacked layers, with each layer being referred to as a ply (see *Figure 8.3a*). In turn, each ply is manufactured by reinforcing a matrix (a polymer or metallic material) with fibers (dispersed or oriented). For an illustrative purpose, *Figure 8.3b* and *Figure 8.3c* depict two plies with 0^0 and 90^0 fiber orientations used to form the stacked laminate in *Figure 8.3a*. Note that, oftentimes, for better performances, the fibers may be arranged based on other angles of interest. Typically, this angle is with respect to the principal material direction, which is the X axis for the laminates shown:

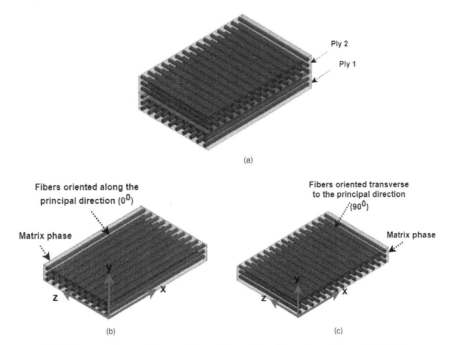

Figure 8.3 – (a) A laminate formed by stacking of two plies, (b) a ply reinforced with fibers arranged parallel to the principal material direction, and (c) a ply reinforced with fibers arranged transversely to the principal material direction

Now, because of the reinforcement with fibers, a single ply may turn out to have either anisotropic or orthotropic material properties. This means their mechanical properties differ along the three X, Y, and Z directions (for anisotropic material) or the variation is restricted to two of the directions for orthotropic. Some important consequences of this anisotropic/orthotropic material behavior are (i) non-uniform distribution of stress across the laminate's thickness, (ii) development of shear stress between the plies – interlaminar shear stress, and (iii) non-suitability of the simple failure theories we have used in the past chapters for the analysis of these materials.

With the background provided, we shall take on a case study to demonstrate the use of SOLIDWORKS Simulation to analyze the response of an engineering component designed with composite material properties.

An analysis of composite beams

This section details our first case study, which is related to the analysis of a curved composite beam acting as an elastic spring.

Through this case study, you will also become familiar with how to create a custom orthotropic composite material and assign its properties to a collection of composite plies. Furthermore, you will become familiar with the effect of fiber orientations on the behavior of composite structures. Moreover, you will learn the difference between the types of stress available for analysis with composite materials and the type of failure criteria used to estimate their factor of safety. Also covered briefly is the procedure to modify SOLIDWORKS Simulation global settings.

Let's get started.

Problem statement

Figure 8.4a shows a multipurpose prosthetic designed with a fiber-reinforced composite [5]. The structural component of the prosthetic consists of two major segments, BA and AC, as highlighted in *Figure 8.4a*. Our focus is on the AB segment (isolated in *Figure 8.4b*) with a thickness of 8 mm and a width of 120 mm. This segment is formed by a curved beam and acts functionally as an elastic spring:

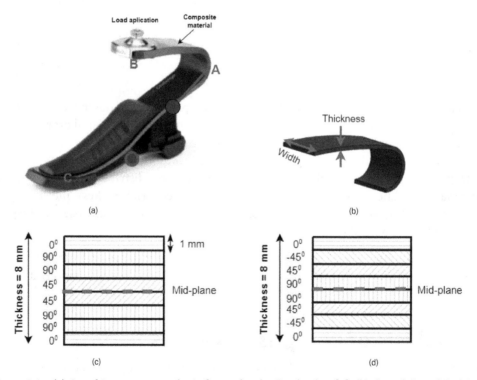

Figure 8.4 – (a) A multi-purpose prosthetic fitness foot by Ottobockus [5], (b) the solid model of the upper elastic spring, (c) a laminate with fiber orientations [0/90₂/45/45/90₂/0], and (d) a laminate with ply orientation [0/-45/45/90/90/45/-45/0]

The goal of the simulation is to investigate the static performance of the BA segment if it were to be designed with a laminate made of **Carbon Fiber-Reinforced Polymer (CFRP)** composites under the combinations of ply orientations shown in *Figure 8.4c* and *Figure 8.4d*. The orthotropic material properties of the CFRP are listed in *Table 8.1*. To be more specific, we want to answer the following questions about the orientations:

- What is the maximum resultant displacement of the elastic spring?
- What is the maximum normal stress across all the plies?
- What is the minimum factor of safety of the structure?

A load of 100 N is assumed to be applied on a small area around point B in *Figure 8.4* of the elastic spring. This area is shown later in *Figure 8.5*:

Properties	Values
Mass density	1,490 kg/m^3
Elastic modulus in X	121 GPa
Elastic modulus in Y	8.6 GPa
Elastic modulus in Z	8.6 GPa
Poisson's ratio in XY	0.27
Poisson's ratio in YZ	0.4
Poisson's ratio in XZ	0.27
Shear modulus in XY	4.7 GPa
Shear modulus in YZ	3.1 GPa
Shear modulus in XZ	4.7 GPa
Tensile strength in X	2,231 MPa
Tensile strength in Y	29 MPa
Compressive strength in X	1,082 MPa
Compressive strength in Y	100 MPa
Shear strength in XY	60 MPa
Yield strength	200 MPa

Table 8.1 – The elastic orthotropic properties of CFRP

As you can see from *Table 8.1*, as opposed to the materials we have seen in all previous chapters, the key mechanical properties for the static analysis of the CFRP composite are specified along the *X*, *Y*, and *Z* directions.

> **Note on Laminate Nomenclature**
>
> In the caption for *Figure 8.4c*, we specified the set of angles for the laminate as $[0/90_2/45/45/90_2/0]$. Note that 90_2 is shorthand for two repetitions of 90. Another shorthand is also commonly used in a composite description when the laminate has symmetric plies. For instance, the two laminates in *Figure 8.4c* and *Figure 8.4c* are symmetric about the midplane. Owing to this, we could have written $[0/90_2/45/45/90_2/0]$ as $[0/90/90/45]_s$, where the s subscript denotes that we have four more plies with a set of angles that mirror the four spelled out, which is $[45/90/90/0]$. In the same spirit, we could also have written $[0/-45/45/90]_s$ for the laminate in the caption for *Figure 8.4d*. You will see the symmetric notation later in this chapter.

In pursuit of the solution to the problem, we shall now go through a series of steps in the upcoming sections to answer the questions attached to the problem statement.

Part A – reviewing the structural model

To get on track with the simulation study, we will first review the model and then activate the simulation study in this section.

Reviewing the file of the curved elastic spring

To get started, download the `Chapter 8` folder from this book's website to retrieve the model of the curved beam. After downloading the folder, follow these steps:

1. You should check to see that it comprises a SOLIDWORKS part file named `Prosthetics_ElasticSpring`.

2. Extract the file and save it on your PC/laptop.

    Let's now briefly go over some of the features of this file. To begin with, you need to get into the SOLIDWORKS environment.

3. Start up SOLIDWORKS.

4. Select **File → Open**, and then open the part file named `Prosthetics_ElasticSpring`.

5. Navigate to the **Feature Manager** design tree and examine the details, as shown in *Figure 8.5*:

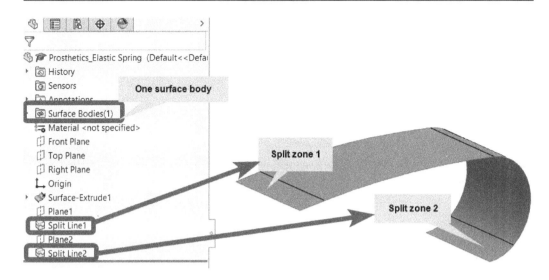

Figure 8.5 – Reviewing the file of the structure

The first thing you will probably notice from *Figure 8.5* is that we have *1* surface body created using the **Surface Extrude** tool. In effect, this surface body represents the surface equivalent of the BA segment in *Figure 8.4b*.

From this, we need to get an important lesson out of the way – *to analyze a composite part within the SOLIDWORKS Simulation environment, you can only employ a surface body (a weldment and solid extruded body are not allowed).*

The other thing you will observe is that there are two split lines (**Split Line1** and **Split Line2**). The split lines are created so that we have separate regions for the application of the load (on zone one) and the application of the boundary condition (on zone two).

Activating the simulation tab and creating a new study

With the key features of the file reviewed, you can go ahead and launch the simulation study as we normally do, as highlighted next:

1. Under the **Command manager** tab, click on **SOLIDWORKS Add-Ins**, and then click on **SOLIDWORKS Simulation** to activate the **Simulation** tab.

2. With the simulation tab active, create a new study by clicking on **New Study**.

3. Input a study name within the **Name** box, for example, `Prosthetics_ElasticSpring Analysis`.

4. Keep the **Static** analysis option, and then click **OK**.

Now that we have activated the simulation study, we can begin to work with the items in the Simulation study tree, which, as shown in *Figure 8.6*, indicates that we need to define the thickness for the surface body:

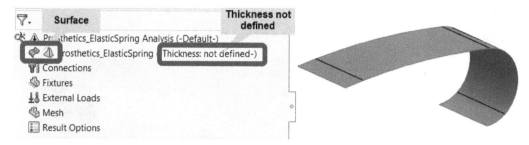

Figure 8.6 – Examining the surface body within the simulation study tree

The next section will walk you through the crucial procedure to convert the surface body into a composite shell. It further discusses some key concepts around the creation of orthotropic composite plies.

Part B – defining the laminated composite shell properties

This section is devoted to the procedure for defining and scrutinizing the properties of the composite plies used in the build-up of the laminated composite shell. It also includes brief subsections on the specification of the fixture and external load on the structure.

Let's begin with the composite material details.

Defining the properties of a multilayered composite shell

The surface body we examined in the previous section will form the basis of the composite shell. In what follows, we shall go through a somewhat long but important series of steps to define the properties of the laminated composite shell.

Carefully follow along:

1. Within the simulation study tree, right-click on the surface body, and then select **Edit Definition**, as shown in *Figure 8.7*:

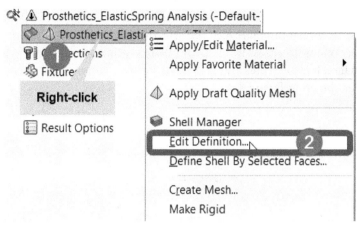

Figure 8.7 – Initiating the thickness definition for the surface body

With this, the **Shell Definition** property manager will pop up, as shown in
Figure 8.8. This is a familiar window that you will have seen in past chapters.
However, we shall not be using the **Thin** option as in all our previous encounters
with this window:

Figure 8.8 – Shell Definition property manager

2. Within the **Shell Definition** property manager, under the **Type** option, select **Composite**, as shown in *Figure 8.8*. The selection of the **Composite** option here immediately opens up access to the **Composite Options** section in the lower region of the manager. It also creates and attaches a separate coordinate system to the shell structure, as shown in *Figure 8.9*:

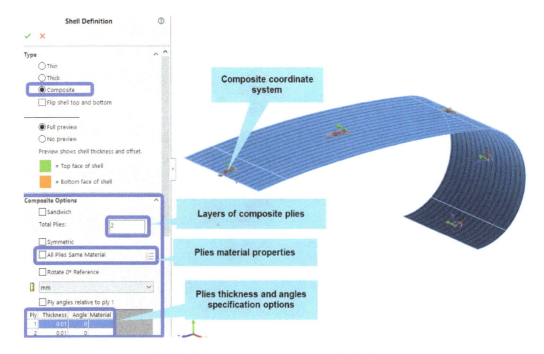

Figure 8.9 – Shell Definition property manager with the Composite option

Note that the gray arrow in the composite coordinate system, which is attached to the shell in the graphics window, always aligns with the ply angle. In other words, it represents the local fiber orientation. Thee items that we shall work with have also been annotated, that is, **Total Plies, Thickness/Angle** box, and the **Material** box. We shall now carry out some further steps with these items.

3. In the **Total Plies** value box (labeled 1 in *Figure 8.10*), key in 8:

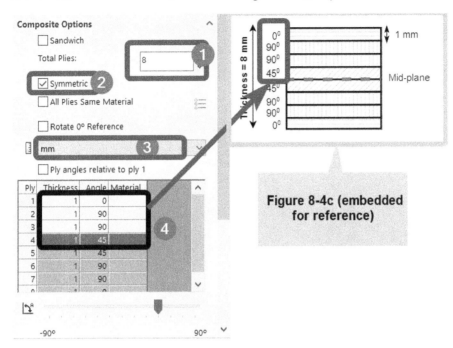

Figure 8.10 – Specifying the ply number, thickness, and angle

4. Tick inside the box labeled 2 (**Symmetric**) and ensure the unit is in **mm** (as indicated in the box labeled 3 in *Figure 8.10*).

5. In the ply property specification box, labeled 4 in *Figure 8.10*, key in, one by one, the four respective values of thickness (1, 1, 1, and 1) and angles (0, 90, 90, and 45).

Note that the values we have keyed in belong to the top half of the composite laminate shown in *Figure 8.4c*. Furthermore, because the laminate is symmetric and we've ticked **Symmetric** in step 4, SOLIDWORKS automatically completes the last four rows in the grayed-out area. It goes without saying that if the ply combination is not symmetric, then we have to add the values of thickness/angle for all plies ourselves.

Now, you will notice that the **Material** column in the box labeled 4 in *Figure 8.10* is empty. Our next task is to define this.

6. To define the material property for the plies, tick inside **All Plies Same Material**, and then click the box labeled 2 in *Figure 8.11*. Doing this will activate the material database:

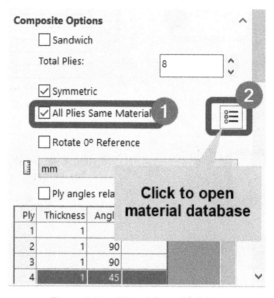

Figure 8.11 – Material specification

7. Within the material database, navigate to the folder named Custom Materials, right-click, and then click on **New Category** (*Figure 8.12a*). With this, a new material folder will be created with a default name of New Category.

8. Change the name to MyComposite (*Figure 8.12b*):

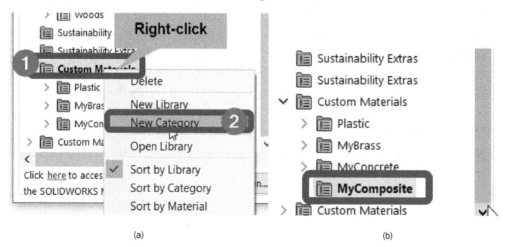

(a)   (b)

Figure 8.12 – Creating a new composite material category under the Custom Materials folder

9. Right-click on the MyComposite folder and click on **New Material**, as shown in *Figure 8.13a*. With this step, a new material file will be created with a default name of Default; modify the name to CustomCFRP, as shown in *Figure 8.13b*:

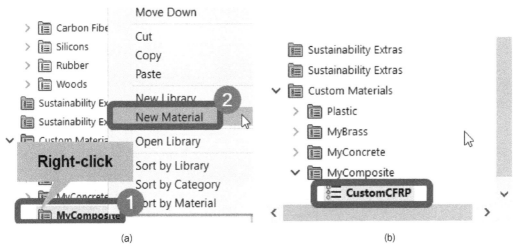

(a)   (b)

Figure 8.13 – Creating the CFRP material file

We shall now work on the right side of the material database to define the properties.

10. On the right side of the material database, under **Model Type**, select **Linear Elastic Orthotropic** (labeled 2 in *Figure 8.14*). This is very important. The default is Linear Elastic Isotropic. Thus, you need to make sure **Linear Elastic Orthotropic** is chosen.

11. Ensure the unit remains as **SI – N/m^2 (Pa)** (labeled 3):

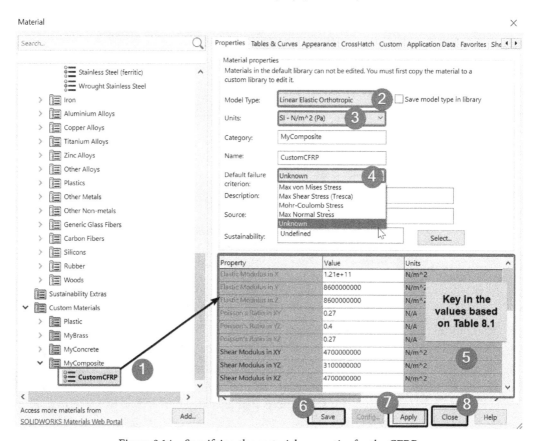

Figure 8.14 – Specifying the material properties for the CFRP

12. Referring to the **Default failure criterion** box (labeled 4 in *Figure 8.14*), choose the **Unknown** option. Now, although we've chosen **Unknown** here, you will see later in the *Obtaining the factor of safety* subsection that SOLIDWORKS will provide the correct list of failure theories for composite materials.

13. For the material property values, enter the values given in *Table 8.1* into the corresponding cells in the box labeled 5 in *Figure 8.14*. Note that only a partial view of the box labeled 5 is shown. You will have to key in 16 property values, as listed in *Table 8.1*.

14. Click **Save**, **Apply**, and **Close** (in that order).

By executing step 14, the material database will close and you will be returned to the **Shell Definition** window, which should now appear as depicted partially in *Figure 8.15*.

As you can see, the material of each ply has been specified as **CustomCFRP**, which we created:

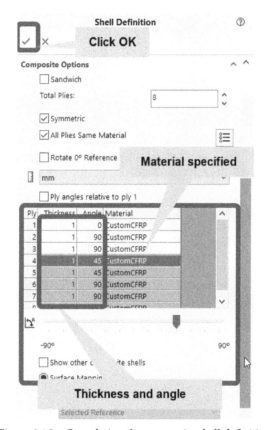

Figure 8.15 – Completing the composite shell definition

15. Click **OK** so that you can return to the simulation window.

Congratulations! You have now fully defined the properties of the composite shell in steps 1–15. With the steps completed, SOLIDWORKS will return you to the usual simulation environment. There, you won't see much difference with the surface body in the graphics window.

However, you will notice that the name of the surface body has a green tick mark on it. Also, if you hover over the name in the simulation tree, it will reveal a clue in the form of **8 / Composite / Middle / CustomCFRP**, as shown in *Figure 8.16*. What does this mean?

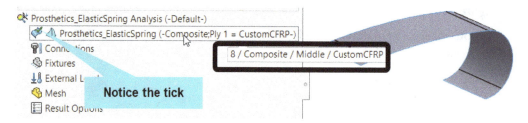

Figure 8.16 – Hovering over the surface body in the simulation tree

Simply put, **8 / Composite / Middle / CustomCFRP** implies that we have 8 plies forming a laminated composite shell derived from the custom CFRP material that we created. Now, you may be wondering why we have the **Middle** item in there? To answer this question, and to further reveal other important details of the laminated composite shell we have defined, we shall now re-scrutinize the shell further in the next subsection.

Scrutinizing the composite shell

To examine some features of the laminated composite shell, follow these steps:

1. Within the Simulation study tree, right-click on the surface body, and then select **Edit Definition** (as we did previously and shown in *Figure 8.7*). This opens up the **Shell Definition** property manager again.

2. Under **Type**, ensure the **Full preview** option is selected (*Figure 8.17a*). The shell will likely appear as shown in *Figure 8.17b*:

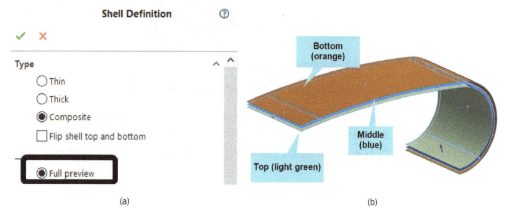

Figure 8.17 – (a) The Full preview option and (b) revealing the shell top, bottom, and middle

What immediately stands out from *Figure 8.17* is that the shell's top and bottom faces are reversed. The light green color denotes the top face. Also, you will also notice that the shell is formed from the surface body by a middle offset method (which is why the ply in blue stays in the middle). We shall correct the position of the top and bottom faces in the next step.

3.   Ensure the **Full preview** option is selected, and then tick inside **Flip shell top and bottom** (labeled 2 in *Figure 8.18a*). This flips the faces, as reflected in the graphics window:

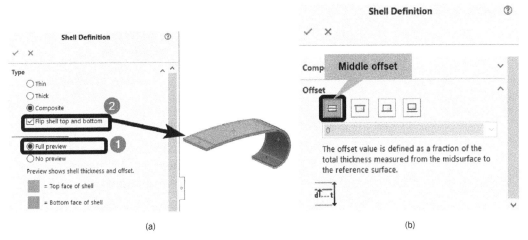

(a)   (b)

Figure 8.18 – (a) Flipping the shell's top and bottom faces, and (b) revealing the Offset option

4.   Next, to confirm the type of offset used to create the shell, scroll to the lower region of the **Shell Definition** property manager, and then expand the item named **Offset**. You will see that the middle offset option is depressed, as shown in *Figure 8.18b*. This step serves only as confirmation of this option. Don't make any changes, as the middle offset approach is often the preferred option to create shells. The other options may be used in cases where the middle offset option will lead to interference of a shell body with another body.

Now that we have seen the offset and flipped the shell appropriately, let's examine another important feature in the next steps.

5.   Scroll to the top of the **Shell Definition** manager, under **Type**, and select the **No preview** option. This will hide the top and bottom faces, and we can then see the fiber directions.

6. Within the **Ply definition** box, click on the first ply and note the update of the shell in the graphics window. You will see the fiber orientation and a coordinate system, as shown in *Figure 8.19*:

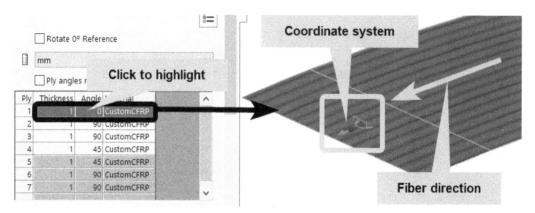

Figure 8.19 – Revealing the fiber direction for ply one

7. Next, click on the **second** and the **fourth** plies, one at a time. Each time, you will observe the shell is updated in the graphics window to show the fiber orientation, as shown in *Figure 8.20a* and *Figure 8.20b* for ply 2 and ply 4 respectively:

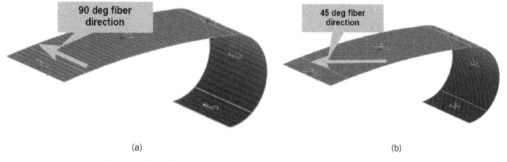

(a)   (b)

Figure 8.20 – Revealing the fiber directions for plies two and four

8. Click **OK** to wrap up the examination.

At this point, we are done with the definitions of the composite shell properties. Specifically, we have created eight plies, each 1 mm thick and all made of CFRP. Note that the eight plies yield a total of 8 mm thickness for the overall composite laminate representing the elastic spring, which is the same as the dimension in *Figure 8.4c*. We have also examined some of the features and options that come with the definition of laminated composite shells. With this, we have provided a sufficient background to move ahead. Therefore, we will now shift our attention to the more familiar steps of applying the fixture and external load.

Applying the fixture and external load

The fixture for this case study involves the application of a fixed boundary condition on zone two of the surface. This zone is located at the bottom of the surface, as shown in *Figure 8.5*.

The steps to apply the fixture are outlined as follows:

1. In the Simulation study tree, right-click on **Fixtures** and pick **Fixed Geometry** from the context menu that appears.

2. When the **Fixture** property manager appears, navigate to the graphics window and click on zone two, as shown in *Figure 8.21*.

3. Click **OK** to complete the fixture specification:

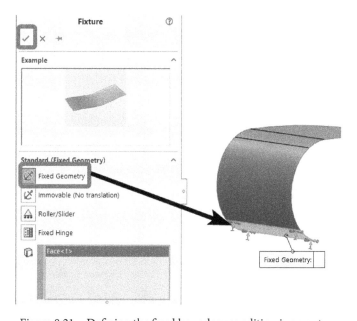

Figure 8.21 – Defining the fixed boundary condition in zone two

Now that we've completed the application of the fixture, let's move on to the application of the external load. This involves the application of a 100 N weight on the area designated as zone one in *Figure 8.5*.

The steps to achieve the application of the load are as follows:

1. Right-click on **External Loads**, select **Force**, and then follow the options indicated in *Figure 8.22*:

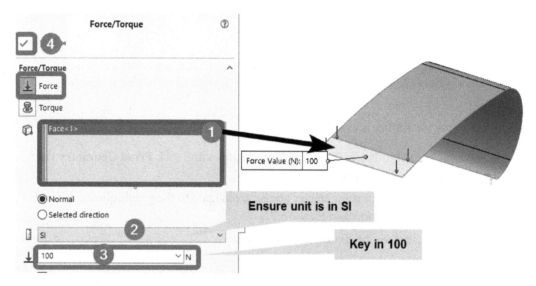

Figure 8.22 – Options for the application of the force on zone one

2. Click **OK** to complete the application of the load.

At this point, we have established how to create and apply composite shell properties to a basic surface body. In effect, this involves the creation of a custom material with *linear elastic orthotropic properties* from scratch. We have also demonstrated how to assign the ply angles and thicknesses for the layers of a laminated composite. Moreover, we've completed the application of the fixture and the external load.

It is now time to get into the meshing phase of the simulation workflow.

Part C – meshing and modifying the system's options

By now, you have seen various ways to carry out the meshing of a structure. So, while this section briefly features the steps for the meshing, it also covers a small subsection that outlines the procedure to modify SOLIDWORKS Simulation's global options so that results can be displayed in the most preferable format later:

Meshing

The meshing strategy we shall adopt is to create the mesh directly without using mesh control. This is because the structure we have for this case study can be considered a simple structure that will be meshed with a single family of **shell elements**. For this reason, let's walk through the meshing as follows:

1. Right-click on **Mesh** and select **Create Mesh**.

2. From the **Mesh** property manager that appears, under **Mesh Density**, drag the **Mesh Factor** to the right, as shown in *Figure 8.23*:

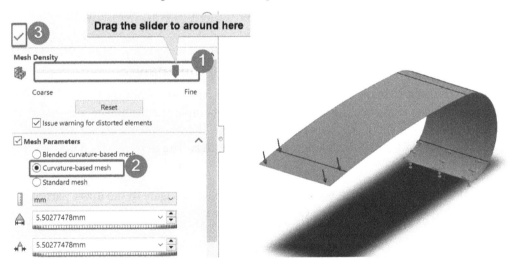

Figure 8.23 – The mesh setting for the structure

3. Under **Mesh Parameters**, change the choice to **Curvature-based mesh**.

4. Click **OK** to complete the meshing actions.

The outcome of the preceding steps is the meshed structure portrayed in *Figure 8.24a*. After completing steps 1–4, you should check the mesh detail by right-clicking on **Mesh** and then selecting **Details** (the result of the mesh detail is shown in *Figure 8.24b*):

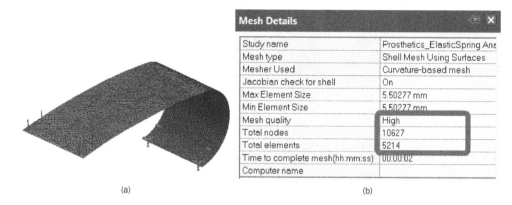

Mesh Details	
Study name	Prosthetics_ElasticSpring Ana
Mesh type	Shell Mesh Using Surfaces
Mesher Used	Curvature-based mesh
Jacobian check for shell	On
Max Element Size	5.50277 mm
Min Element Size	5.50277 mm
Mesh quality	High
Total nodes	10627
Total elements	5214
Time to complete mesh(hh:mm:ss)	00:00:02
Computer name	

(a)   (b)

Figure 8.24 – (a) The meshed structure and (b) mesh details

This completes the meshing phase, which leaves us with the next set of actions geared toward modifying the simulation's global options.

Modifying the system options and running the analysis

Ordinarily, we would have dived straight into running the analysis, but the post-processing of results when dealing with an analysis that employs a composite material can get overwhelming. For this purpose, we shall streamline the way the results are displayed by modifying SOLIDWORKS Simulation's systems settings. The steps required for this are highlighted next:

1. From the main menu, click on **Simulation**, and then navigate to the lower part of the pull-down menu to click on **Options**, as shown in *Figure 8.25*:

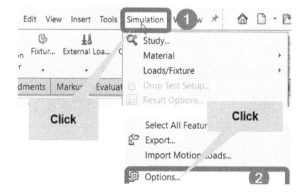

Figure 8.25 – Activating the simulation's options

2. From the **Options** property manager that pops up, navigate to the **Default Options** tab.

3. Click on **Color Chart** (labeled 2 in *Figure 8.26*).

4. Move the to the right side of the window, under **Number format**; change this to **Floating** (labeled 3 in *Figure 8.26*).

5. Click **OK**:

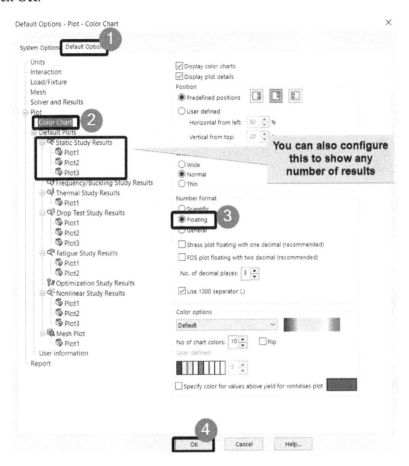

Figure 8.26 – SOLIDWORKS Simulation's system options

Many more options can be modified under the **System Options** and **Default Options** tabs. However, for the sake of brevity, we shall stick to this minor but important change. It is recommended that you look carefully at the various options; they will likely come in handy in your future explorations. In contrast to what we have done in the past chapters by accepting the default option, the change enacted in the preceding steps will save us the need to have to change the number format every time we retrieve a result.

With this, let's now transition to the running and retrieval of the results.

Part D – the running and post-processing of results

Overall, there are five subsections here to walk you through the running of the analysis and post-processing of the results.

Running the analysis and examining the default resultant displacement

Follow these steps to launch the solver:

1. Right-click on the study name and then select **Run**. After the running is complete, under the Results folder within the Simulation study tree, you will see the usual three default results (**Stress 1**, **Displacement1**, and **Strain1**).

2. Double-click on **Displacement1 (-Res disp-)** to display its plot in the graphics window, as shown in *Figure 8.27*:

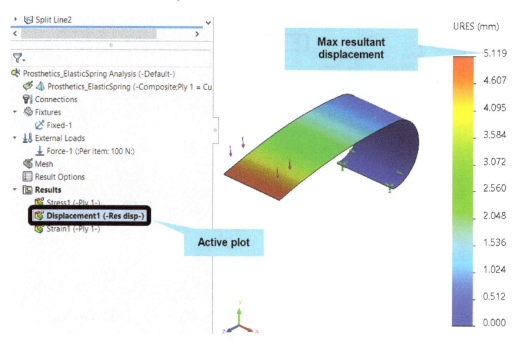

Figure 8.27 – The plot of the resultant displacement

The plot in *Figure 8.27* indicates that the maximum value of the resultant displacement is *5.119 mm*. You may be puzzled by why we are getting the resulting displacement here rather than the vertical displacement along the *Y* axis. For this study, we are using a collection of **shell elements** to investigate the response of a composite-based component. Now, due to the variation in the fiber orientations of each ply in the laminated composite, the nodes of the discretized body are bound to experience complex movements along the *X*, *Y*, and *Z* directions. Consequently, the resultant displacement is a better indicator of the cumulative deformation of the composite structure than the vertical displacement along the *Y* axis (even though the load is normally applied solely to the surface in the *Y* direction). This is why it has been selected here. Meanwhile, take note that whatever values we obtain in this subsection belong to the composite laminate with the [0/90/90/45]s configuration shown in *Figure 8.4c*. We will later see in the *Comparing displacements and stresses* subsection how this compares with the second configuration in *Figure 8.4d*.

With the maximum resultant displacement value noted, we next aim to address the following questions:

- What is the maximum normal stress across the plies?
- What is the distribution of the factor of safety?

Obtaining the stress results

The following steps should be followed to obtain the maximum normal stress across the plies:

1. Right-click on the `Results` folder and then select **Define Stress plot**.

2. The **Stress plot** property manager should appear, as shown in *Figure 8.28a*:

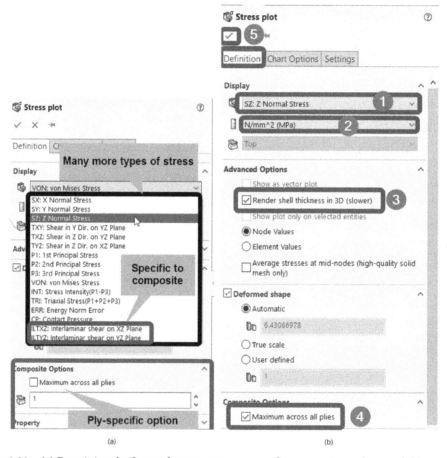

(a)   (b)

Figure 8.28 – (a) Examining the Stress plot property manager for composite analysis and (b) specifying options for the normal stress along the Z direction (major material direction)

As you can see from *Figure 8.28a*, this **Stress plot** manager differs in appearance from what we have seen in the past chapters. The first difference is some newly available stress options that are specific to composite analysis, specifically the interlaminar shear stresses. Another notable difference is the option to configure the stress assessment for a specific ply, as shown under **Composite Options** in *Figure 8.28a*.

3.   For the focus of this study, within the **Stress plot** property manager, under **Display**, follow the options highlighted in *Figure 8.28b* to retrieve the **Maximum across all plies** normal stress along the principal material direction. Note that you can iteratively examine the stress for each ply. However, for brevity's sake, we will not pursue this option. But it is recommended that you explore this further.

4.   Click **OK**.

With the foregoing steps, the graphics window is updated with the plot of the normal stress along the *Z* direction, as shown in *Figure 8.29*:

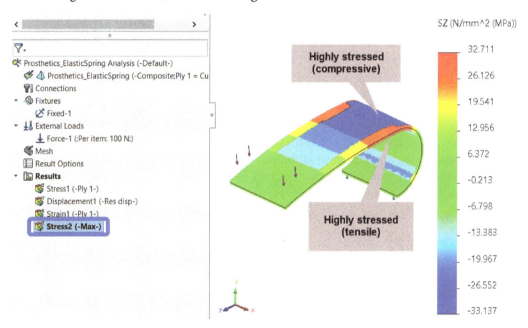

Figure 8.29 – A plot of the normal stress along the Z axis (maximum across all plies)

From this plot, we can infer that the elastic spring experiences a maximum tensile normal stress across all plies of *32.711 MPa* and a maximum compressive normal stress across all plies of *-33.137 MPa*. These values are relatively small compared to the strengths of the composite, which means the spring can sustain even more weight than the specified value of 100 N.

So far, we have retrieved the maximum resultant displacement and the maximum normal stress across all plies. Notably, the values we've obtained for these two important engineering design variables are not too excessive to raise concerns about the safety of the structure. Nonetheless, for a broader perspective on the performance of the design, we shall next examine the factor of safety.

Obtaining the factor of safety

Keep in mind that we discussed the **Factor of Safety** (**FOS**) in *Chapter 2, Analyses of Bars and Trusses*. There, we looked at the concept in the context of isotropic material. In contrast, for this study, we are dealing with an orthotropic fiber-reinforced composite material, which belongs to a family of materials with complex failure mechanisms. Nonetheless, irrespective of the material type used in calculating the FOS, two things should be clear:

- The FOS must be based on a certain failure criterion. As you will soon see, SOLIDWORKS Simulation provides three types of failure criteria for composite materials. These are the *Tsai-Hill* criterion, the *Tsai-Wu* failure criterion, and the maximum stress criterion. In what follows, we shall use the Tsai-Wu criterion. For more detailed coverage of the mathematical formulation behind these criteria, the excellent book by *Hyer and White [4]* is recommended.

- The greater the FOS is (than 1), the better the safety rating is of the component (all things being equal).

Now, to obtain the FOS, follow these steps:

1. Right-click on the `Results` folder and then select **Define Factor of Safety Plot**.
2. The **Factor of Safety** property manager appears, as shown in *Figure 8.30a*.
3. Under **Composite Options**, select **Tsai-Wu Criterion**, as indicated in *Figure 8.30a*:

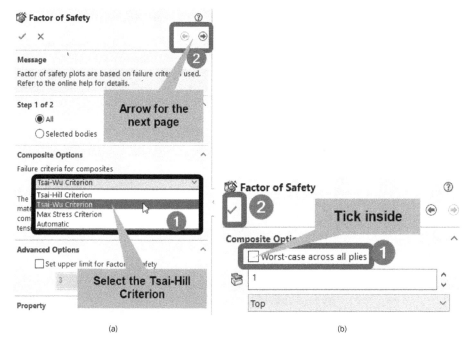

Figure 8.30 – (a) The first page of the Factor of Safety property manager and (b) The second page of the Factor of Safety property manager

4. Click on the **Next** arrow (labeled 2 in *Figure 8.30a*) to move to the second page of the property manager, which should appear as shown in *Figure 8.30b*.

5. Check the **Worst-case across all plies** box (labeled 1 in *Figure 8.30b*).

6. Click **OK** to exit the **Factor of Safety** property manager.

The plot of the factor of safety will be displayed in the graphics window.

7.   With the plot displayed, employ the **Probe** tool, as shown in *Figure 8.31*, and then navigate to the graphics window to pick four locations to assess the variation of the FOS:

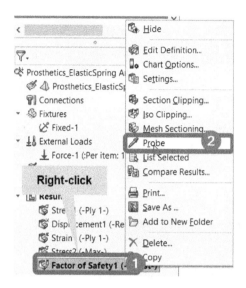

Figure 8.31 – Using the Probe tool with the FOS

The final result of the FOS plot with four locations that have been picked is shown in *Figure 8.32*:

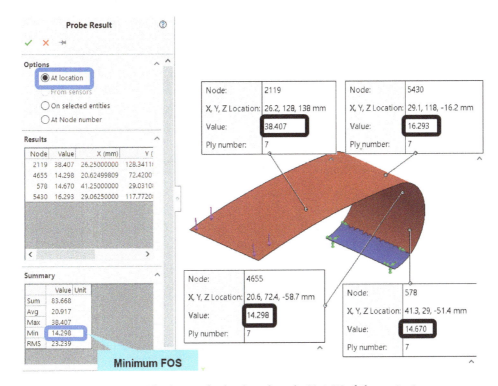

Figure 8.32 – The factor of safety based on the Tsai-Wu failure criterion

The summary table in the lower-left corner of *Figure 8.32* indicates that the structure sustains a minimum FOS of roughly 14. This is a strong indication that the design is technically safe.

At this point, we have been able to address the questions that we set out to answer at the beginning of the problem about the values of the maximum resultant displacement, the maximum normal stress along the principal direction, and the minimum factor of safety. However, these values have only been obtained for the $[0/90/90/45]_s$ ply orientation of *Figure 8.4c*. For the sake of completeness, we will now briefly reckon with the second ply orientation (*Figure 8.4d*) in the next subsection.

Examining the response of the laminate with the ply set of [0/-45/45/90]$_s$

The major thing we need to cover in this subsection is to determine the values of the resultant displacement and the maximum normal stress obtained for a laminate with the [0/-45/45/90]s orientation. This will allow us to determine whether there is any significant difference between the ply orientations.

Now, to analyze the new laminate orientation, we will duplicate the study we have completed in the previous sections so that we don't have to start the study from scratch.

To duplicate the study, follow the steps given next:

1. Navigate to the base of the graphics window, right-click on the study tab, and then select **Copy Study**, as shown in *Figure 8.33a*.

2. Within the **Copy Study** property manager that pops up, provide a name for the new study, for example, Prosthetics_ElasticSpring-Laminate2, as indicated in *Figure 8.33b*.

3. Click **OK**:

Figure 8.33 – (a) Duplicating the completed study and (b) naming the new study

After completing the preceding steps, a new study tab will be launched beside the old study tab. Ensure you remain in this new study environment. With the study copied, the next action is to edit the shell to reflect the new laminate configuration. The following steps show you how to edit the shell for the new study:

1. Within the Simulation study tree, right-click on the surface body and then select **Edit Definition** to open up the **Shell Definition** property manager (for a guide, you can refer to *Figure 8.7*).

2. Under **Type**, select the **No preview** option (if need be, see *Figure 8.17a*).

3. Navigate to the **Ply definition** box to update the ply angles, as shown in *Figure 8.34*. Note that only the angles of the top four plies need to be changed. It is recommended that you observe the fiber orientation on the shell as you modify the angles:

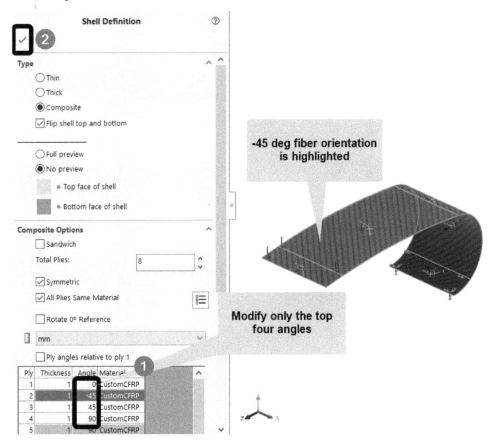

Figure 8.34 – Updating the ply angles for the new laminate

4. Click **OK** to wrap up the update of the ply angles.

5. Next, run the analysis by following *Figure 8.35*:

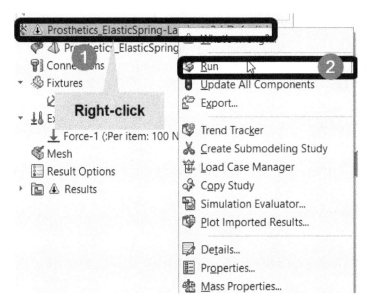

Figure 8.35 – Running the analysis with the new laminate set

After the running process is completed, we will have a new set of results. The key task for us now is to compare the results.

Comparing displacements and stresses

To compare the results of the analyses conducted with the previous laminate ([[0/90/90/45] s) and the new laminate ([0/-45/45/90]s), follow these steps:

1. Right-click on the Results folder in the current analysis.

2. Select **Compare Results** (*Figure 8.36a*).

3. Within the **Compare Results** property window that appears, under **Options**, tick inside **All studies in this configuration** (labeled 1 in *Figure 8.36b*).

4. To compare the values of the resultant displacements from the two studies, make the selection as depicted in the box labeled 2 in *Figure 8.36b*.

5. Click **OK** to complete the selections for the comparison:

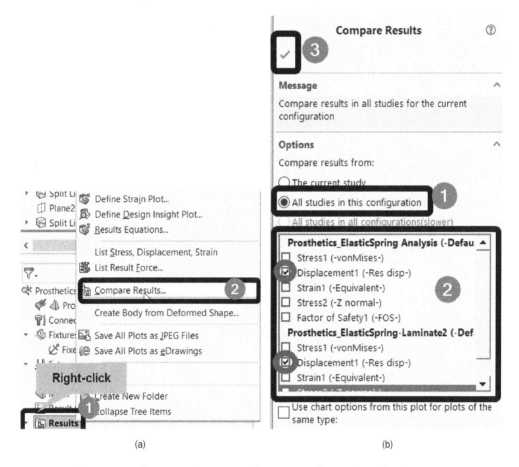

(a)   (b)

Figure 8.36 – Initiating the process of comparing the resultant displacements

Upon completing the preceding actions, the graphics window will be updated to show the plots of the resultant displacements for the two studies. This is shown in *Figure 8.37*:

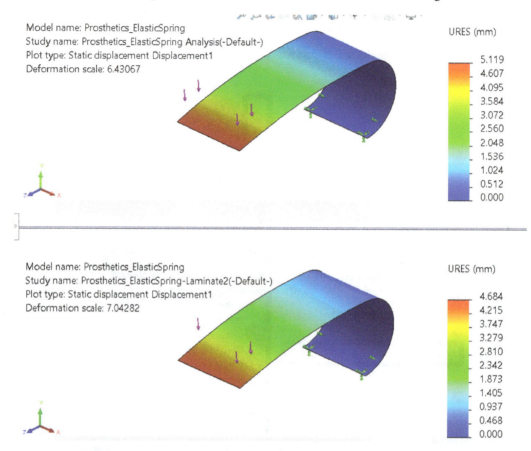

Figure 8.37 – The plots of the resultant displacements for the two studies

As you can see from *Figure 8.37*, the maximum resultant displacement for the laminate in the first study is *5.119 mm*, while the maximum resultant displacement for the laminate in the second study is *4.684 mm*. This difference amounts to a 9% lower deformation exhibited by the laminate with the ply set of $[0/-45/45/90]_s$ compared to that of the other laminate.

But can we say the same about the stress in the structure? Luckily, we can obtain a similar plot comparing the normal stress for the two studies. For this, you need to repeat steps 1–5. But instead of picking **Displacement1(-Res disp-)** for both studies (as was depicted in *Figure 8.36b*), pick **Stress2 (-Z normal-)** in step 4. With the steps executed, the outcome is shown in *Figure 8.38* for the comparison of the normal stresses.

From the stress plots, it turns out that there is a relief of the compressive normal stress in absolute term with the [0/-45/45/90]$_s$ laminate, which experiences stress of *31.23 MPa* compared to the *33.137 MPa* that was experienced by the [0/90/90/45]$_s$ laminate. Meanwhile, the reality for the tensile stress is a bit different. The [0/-45/45/90]$_s$ laminate endures about 2% more tensile stress than what was induced in the [0/90/90/45]$_s$ laminate. All in all, given the lower deformation and the lower compressive stress, we can make the case that the [0/-45/45/90]$_s$ laminate offers a somewhat better static performance than the [0/90/90/45]$_s$ laminate.

As a final remark on this, in practical product design scenarios, ease of manufacturing will also feed into the decision to trade off one configuration of a laminate for another or one material for another. Occasionally, it may also be necessary to investigate the fatigue behavior of both laminates in addition to the basic stress/deformation analyses conducted here. We shall look at fatigue analysis in *Chapter 9, Simulation of Components under Thermo-Mechanical and Cyclic Loads*:

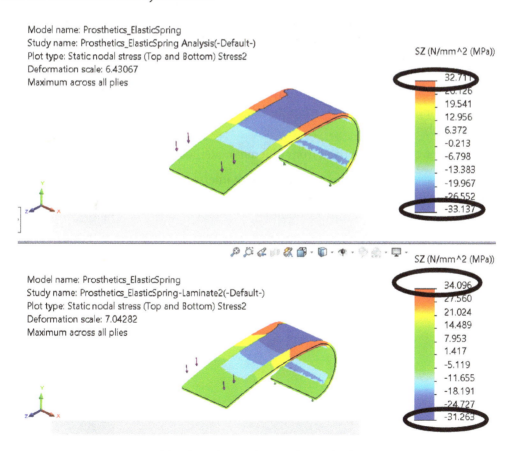

Figure 8.38 – The plots of normal stress along Z for the two studies

This concludes our exploration of the case study. Through this case study, we have touched on important concepts to give you the background to build on concerning the analyses of components with composite materials. Over the many sections, you have seen the creation of a custom orthotropic composite material, the application of a composite material to a laminate, the definition of ply properties, and the visualization of fiber orientations. Also, you have learned the difference between the types of stress and failure criteria available for analyses with composite materials. Finally, we have leveraged the *Compare Results* tool in SOLIDWORKS Simulation to compare the performance of laminates with different sets of ply orientations. In the next section, we shall take a brief look at some other points to be mindful of in the simulation of advanced composite structure.

Analysis of advanced composite structures

The world of analysis with composite materials is vast. So far, in the completed case study, we have dealt with a component constructed of an 8-ply laminate, with all plies having the same material (which is CFRP). For advanced composite structures, things can get pretty complex. For this reason, among other things, you need to know the following:

- In SOLIDWORKS Simulation, you can only have a maximum of 52 plies.

- If you have a total number of plies, say n, take note that SOLIDWORKS's ply-counting procedure starts from the bottom surface (ply 1) to the top surface (ply n).

- You can have each ply derived from different materials, not just one material.

- You can have complex geometric bodies with intersections. Nevertheless, no matter how complex a structure may be, to analyze it with a composite material, it must be created as a collection of surface bodies.

With this achieved, you can leverage the skills and workflow demonstrated in the previous sections to explore the analysis of advanced composite structures.

Summary

We started this chapter with an overview of composite materials as they relate to the analysis of composite structures. We finish having conveyed fundamental concepts for the static analysis of composite structures using the SOLIDWORKS simulation capabilities. Using a composite elastic spring in the case study, we have demonstrated, among other things, the following:

- How to convert a basic surface body into a composite shell

- How to create a custom orthotropic composite material, assign its properties to a composite laminate, and visualize the fiber orientations of composite plies

- How to obtain the stress cross plies and employ the right failure criteria to assess the factor of safety for composite structures

- The benefit of editing the global simulation settings to enhance the presentation of results

By bringing the knowledge of this chapter and the previous chapters together, you have acquired the foundational skill set to be able to handle the various kind of analyses of composite structures. In the next chapter, you will learn about the analysis of components with thermo-mechanical and cyclic loads.

Exercise

1. Re-conduct the case study to compare the performance of the [0/-45/45/90]$_s$ and [45/-45/45/90]$_s$ laminates. By using the same load and boundary condition, determine the percentage difference in the resultant displacements engendered by both laminates. What is the difference in interlaminar stress between the two?

2. Re-conduct the case study to investigate the performance of the elastic spring with a laminate of the [0/90/90/0]$_s$ form. How does it compare with [0/-45/45/0]$_s$ in terms of deformation?

Further reading

- [1] *Materials Selection in Mechanical Design, M. F. Ashby, Elsevier Science, 2016*
- [2] *A parametric study of flutter behavior of a composite wind turbine blade with bend-twist coupling, Composite Structures, Vol. 207, P. Shakya, M. R. Sunny, and D. K. Maiti, pp. 764–775, 2019*
- [3] *Introduction to composite materials, Stability and Vibrations of Thin Walled Composite Structures, H. Abramovich, Elsevier, pp. 1–47, 2017*
- [4] *Stress Analysis of Fiber-reinforced Composite Materials, M. W. Hyer and S. R. White, DEStech Publications, Incorporated, 2009*
- [5] *A multi-purpose prosthetic fitness foot, Ottobock,* `https://www.ottobockus.com/fitness/solution-overview/challenger-foot/,` `(accessed 2021)`

9
Simulation of Components under Thermo-Mechanical and Cyclic Loads

In the preceding chapters, we demonstrated various types of loads in the static analysis of engineering components. Among other things, there are two major ideas implicit in those chapters: (i) the assumption that the component is loaded under normal ambient temperature conditions, and (ii) the assumption that the applied load history is non-cyclical and therefore will not lead to fatigue. But there are many practical cases where these assumptions are violated. The good news is that SOLIDWORKS Simulation is equipped to deal with cases where the violation of these assumptions happens. Thus, breaking away from the aforementioned assumptions, this chapter will focus on the simulation of components under the effects of thermal and cyclical loads. Two case studies will be deployed to illustrate the strategies for dealing with the two effects separately under the following topics:

- Analysis of components under thermo-mechanical loads
- Analysis of components under cyclical loads

Technical requirements

You will need to have access to the SOLIDWORKS software with a SOLIDWORKS Simulation license.

You can find the sample file of the model required for this chapter here:

`https://github.com/PacktPublishing/Practical-Finite-Element-Simulations-with-SOLIDWORKS-2022/tree/main/Chapter09`

Analysis of components under thermo-mechanical loads

This section initiates our entry into the exploration of the SOLIDWORKS Simulation for the analysis of components subjected to a combination of thermal and mechanical loads. The problem we will deploy for this purpose stems from the design of a diaphragm-based pressure sensor.

Problem statement

Suppose you are involved in the prototyping of a differential pressure sensor for the measurement of ultra-low pressure in the range of 0–5 bars (or 0–0.5 MPa). Let's say you have narrowed down the key functional components of the measurement to comprise an enclosed circular diaphragm and a Wheatstone bridge, as shown in *Figure 9.1a*. Principally, it is known that this specific pressure sensor works by converting the deflection of a circular diaphragm into an electrical signal *[1]*, where the deflection of the diaphragm is caused by the difference between a reference pressure (P_{ref}) in an enclosed chamber and an incoming target pressure (P_{tar}), as shown in *Figure 9.1b*:

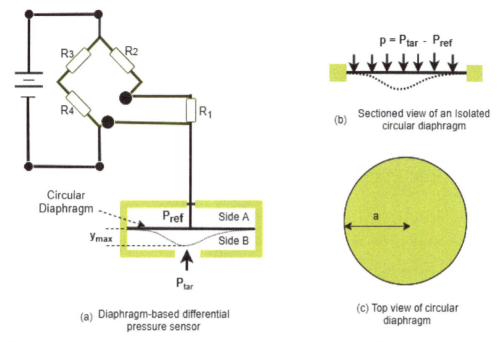

(a) Diaphragm-based differential pressure sensor

(b) Sectioned view of an Isolated circular diaphragm

(c) Top view of circular diaphragm

Figure 9.1 – The schematics of the key components of a diaphragm-based differential pressure sensor

The objective is to design the pressure sensor for use in a hostile environment, which will see side A (see the annotation in *Figure 9.1a*) of the diaphragm loaded with a maximum pressure of 0.5 MPa. Furthermore, it is known that side B may be occasionally exposed to a temperature of 54.84°C arising from the state of the target fluid.

To aid the downstream fabrication of the sensor, an exploratory design analysis is to be conducted to test the performance fidelity of a *chromium stainless steel* diaphragm with a thickness of 3 mm and diameter of 60 mm under the effects of 0.5 MPa pressure. However, while it is expected that this combination of initial dimensions will produce a diaphragm with sufficient static strength, the additional thermal stress generated by the temperature difference at the faces of the diaphragm may lead to failure down the road. For this reason, the goal of this case study is to conduct an integrated static and thermal analysis of the diaphragm.

The solution to the preceding problem is organized into three major sections. By the end of this problem, you will become familiar with the following:

- A procedure to set up a basic thermal analysis and link the analysis with a static analysis

- A strategy to conduct an optimization study to determine the optimal dimensions of the diaphragm for performance against failure

We shall kick things off by reviewing the model provided to aid the simulation study in the next section.

Reviewing the file of the circular diaphragm

To begin the review, download the Chapter 9 folder from the book's website to retrieve the model of the diaphragm. After downloading the folder, do the following:

1. You should check to see that it comprises a SOLIDWORKS part file named Diaphragm.

2. Extract the file, save it on your PC/laptop, and start up SOLIDWORKS.

3. Open the part file named Diaphragm via **File → Open**.

4. Navigate to the **Feature Manager** design tree to examine the details, as shown in *Figure 9.2*:

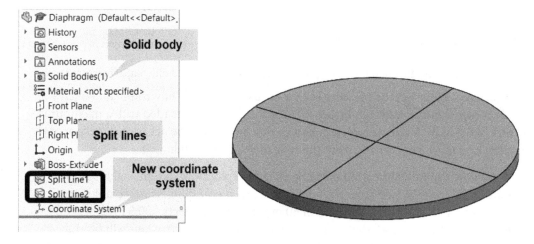

Figure 9.2 – Reviewing the file of the diaphragm

From *Figure 9.2*, you will notice that the model comes with two split lines and one coordinate system. The split lines are needed to facilitate the creation of the new coordinate system. The split line may also be used to examine the variation of radial stress from the center of the diaphragm to the edge.

5. Right-click on **Coordinate System1**, and then select **Edit Feature** to expose the **Coordinate System** property manager, as shown in *Figure 9.3*.

Notice that the coordinate's Y-axis goes through the circumference of the diaphragm. We will come back to this in the later section when extracting the diaphragm's stress results:

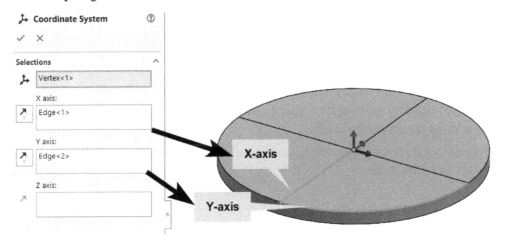

Figure 9.3 – Examining the axis of the coordinate system

6. Click **OK**.

This concludes the review, and with the review out of the way, it is time to launch the study.

Dealing with the thermal study

Analyses that involve heat transfer, which is mostly driven by temperature change, are classified broadly under the name *thermal study* or *thermal analyses*. Thermal analysis has a unique set of vocabulary and it is a broad topic, with dedicated books covering various aspects of its theoretical foundations *[2]*.

In the context of SOLIDWORKS Simulation, the thermal study environment can be used to investigate the following:

- Transient thermal analyses, where the heat transfer process involves rapid changes with time

- Steady-state thermal analyses, under which the parameters of the heat transfer process do not change rapidly with time

With any of these two types of thermal analysis, the key degree of freedom is the temperature (which is analogous to displacement in static analysis). Moreover, for general thermal analysis, the most important properties of the material needed for the simulation are *thermal expansion coefficient*, *thermal conductivity*, and *specific heat capacity*. Materials within SOLIDWORKS's database without these properties cannot be used for thermal analysis unless you add the properties manually.

In what follows, we shall restrict our focus on a steady-state analysis that involves prescribed temperature states.

Creating the thermal study

With this being the first time dealing with a thermal study in this book, it is good to point out that the purpose of this subsection is to factor in the thermal load caused by the temperature difference on the faces of the diaphragm.

Follow these steps to activate the simulation add-in and launch the thermal study:

1. On the **Command manager** tab, click on **SOLIDWORKS Add-Ins**, and then click on **SOLIDWORKS Simulation** to activate the **Simulation** tab.

2. With the **Simulation** tab active, create a new study by clicking on **New Study**.

3. Input a study name within the **Name** box, for example, `Diaphragm Analysis`.

4. Under **Advanced Simulation**, select the **Thermal** analysis option, as shown in *Figure 9.4*:

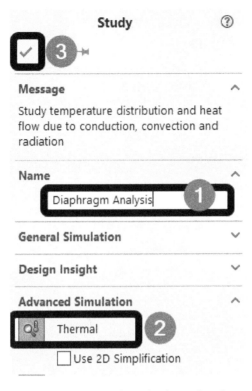

Figure 9.4 – Launching the thermal study

5. Click **OK**.

Now that we have launched the thermal simulation study, we can begin to work with the simulation settings.

Selecting thermal options, assigning material, and specifying temperatures

Here, we are going to specify all the requirements for the thermal analysis. Primarily, we shall define the material for the diaphragm, appropriately assign the right temperature values to its top and bottom faces, and run the analysis. The steps to do this are outlined next:

1. To specify the thermal analysis option, right-click on the analysis name and select **Properties,** as shown in *Figure 9.5a*. This will launch the thermal study dialog box shown in *Figure 9.5b*:

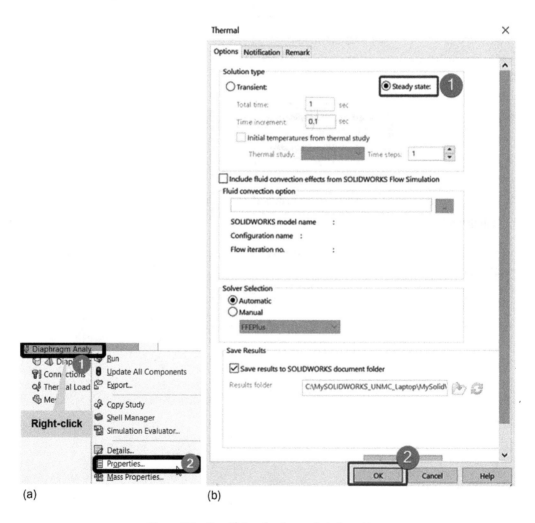

(a)   (b)

Figure 9.5 – Specifying the thermal study option

2. Under the **Options** tab of the thermal study dialog box, ensure the **Steady state** type is selected, as shown in *Figure 9.5b*. This option is consistent with the fact that the temperatures of the surfaces are assumed to remain constant over time.

3. Click **OK** to close the material database.

4. Back in the simulation study tree, right-click on the part's name (Diaphragm) and choose **Apply/Edit Material** (*Figure 9.6a*):

(a)   (b)

Figure 9.6 – Assigning material to the diaphragm

5. From the materials database that appears, expand the Steel folder, and then assign **Chrome Stainless Steel** (as indicated in *Figure 9.6b*).

6. Click **Apply** and and click **Close** to close the material database.

    Steps 1–6 complete the specification of the study option and the assignment of material for the thermal study. We will now move on to the specification of temperature.

7. Right-click on **Thermal Loads** and select **Temperature…**, as shown in *Figure 9.7*:

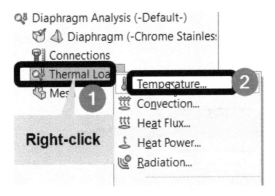

Figure 9.7 – Initiating the application of temperature load

Within the **Temperature** property manager that appears, execute the following actions.

8. Click inside the reference box (labeled 1 in *Figure 9.8*) and navigate to the graphics window to select the top face of the circular plate diaphragm.

9. Ensure the unit of temperature is **Celsius** (labeled 2 in *Figure 9.8*).

10. Specify a temperature of 24.85 (labeled 3 in *Figure 9.8*):

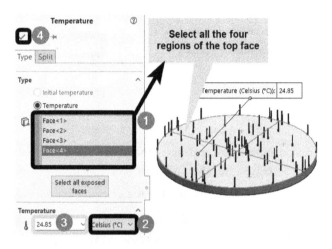

Figure 9.8 – Specifying the temperature for the top face

11. Repeat *step 7*, and then follow *Figure 9.9* to apply a temperature of 54.85 to the bottom of the circular plate diaphragm:

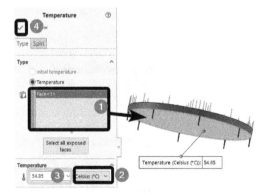

Figure 9.9 – Specifying temperature for the bottom face

12. Mesh and run the thermal analysis by using the **Mesh and Run** command, as shown in *Figure 9.10*:

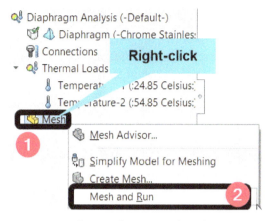

Figure 9.10 – Meshing and running of the thermal analysis

Once the meshing and running operation is complete, the graphics window will display the thermal plot, as shown in *Figure 9.11*. As you can see from *Figure 9.11*, there is only one default result named **Thermal1** in the `Results` folder (unlike in static simulations where three default results are generated after the successful running of a study). This result corresponds to the temperature plot shown in the graphics window. Note how the heat permeates from the bottom to the top side of the diaphragm. Additionally, you will observe that the maximum and minimum temperatures depicted in *Figure 9.11* agree with the temperatures we specified for the bottom and top faces of the diaphragm in *steps 10* and *11* respectively. This agreement is to be expected since we are dealing with a basic steady-state thermal analysis, with prescribed temperature loads and no consideration of convective heat loss:

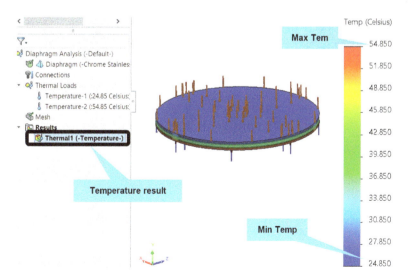

Figure 9.11 – Temperature distribution due to thermal loads on the top and bottom faces

Although we've dealt with a temperature load for this analysis, more complex thermal analyses may involve the application of thermal boundary and forcing functions in the form of *heat flux*, *heat power*, *convection*, and *radiation* transfer processes (see the options below **Temperature…** in *Figure 9.7*). If you want to dig deep into these other options, a good starting point is a book by *Kurowski [3]*.

This concludes the thermal analysis. To reiterate, with the activities carried out in this subsection, we have delved into a steady-state thermal analysis through which we have applied temperatures on the faces of the diaphragm. Technically, what follows from the exposure of the bottom face of the diaphragm to the high temperature is radial expansion. In the next section, we will see how this expansion plays out when the diaphragm is constrained on the edge during the static study.

Dealing with the combined thermal static study

Part of our objective with the static study to be conducted here is to apply external pressure on the surface of the diaphragm so that its maximum deflection can be quantified under a proper constrain. But in addition to pressure load, we shall also carry over the effect of thermal analysis into the static analysis.

Launching the static study

The following steps should be followed to launch the static study:

1. Navigate to the base of the graphics window, right-click on the **Study** tab, and then select **Create New Simulation Study**, as shown in *Figure 9.12a*.

2. Within the **Study** property manager that pops up, provide a name for the new study, for example, Diaphragm Static, as shown in *Figure 9.12b*.

3. Select the **Static** option, as shown in *Figure 9.12b*.

4. Click **OK**.

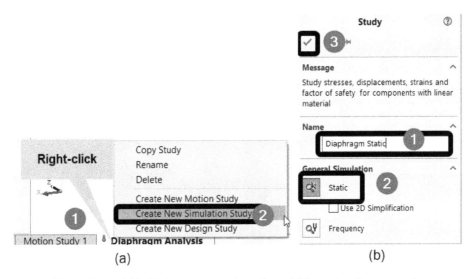

Figure 9.12 – (a) Initiating a new static study and (b) naming the new study

Once *steps 1–4* are completed, a new study tab for the static study will be created beside the thermal study tab at the bottom of the graphics window. With this done, our next focus is to integrate the thermal results with the static study (this is the salient feature of this case study).

Integration of thermal results with static study

In contrast to our previous foray into the static study, we need to incorporate the temperature distribution from the thermal analysis into the static study. To achieve this, follow these steps:

1. Ensure you are within the newly created static study, as shown in *Figure 9.13*. While within the static study window, also take note of the status of the external load (currently empty):

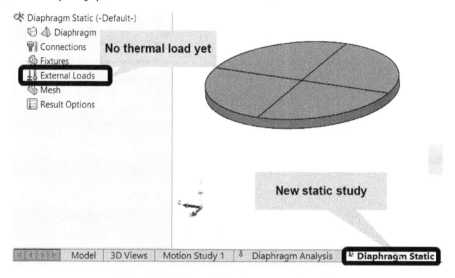

Figure 9.13 – Examining the new static study

2. Right-click on the analysis name, and then select **Properties** (as we did in *Figure 9.5a*).

    The preceding step initiates the static study dialog box within which we now need to make the following changes.

3. Navigate to the **Flow/Thermal Effects** tab, as shown in *Figure 9.14*.

4. Under **Thermal options**, tick inside **Temperatures from thermal study** (labeled 1 in *Figure 9.14*). This step will link the thermal results from the *Dealing with the thermal study* section with the current static study:

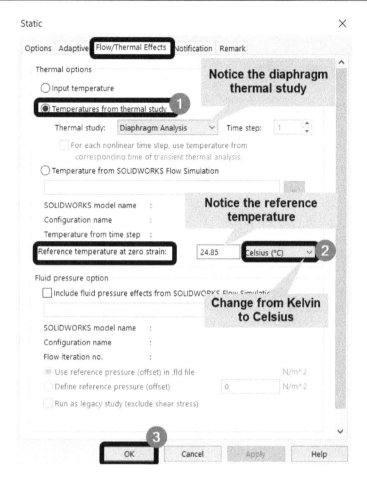

Figure 9.14 – Modifying the static study dialog box to include the thermal effect

5. In the box labeled 2, change the unit of temperature from **Kelvin (K)** to **Celsius (°C)**.

Take note of the number 24.85, which indicates the **Reference temperature at zero strain** value.

6. Click **OK**.

The completion of *steps 1–6* introduces a new item called **Thermal** under **External Loads**, as displayed in *Figure 9.15*:

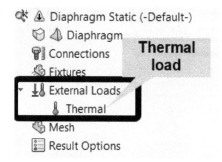

Figure 9.15 – Evidence of a linked thermal load for the static study

---

A Remark on a Common Error in Thermal Stress Analysis

You may have noticed that the value of the *reference temperature at zero strain* (which is 24.85°C) that was highlighted in *step 5* of this subsection matches the value of the temperature we applied in *step 9* of the *Selecting thermal options, assigning material, and specifying temperatures* subsection. This value (24.85°C) is considered to be the ambient temperature at which material properties such as Young's modulus and Poisson's ratio are obtained during experimental testing. Consequently, it means that the top face of the diaphragm is assumed to be at ambient temperature, while its bottom face, which is exposed to a temperature of 54.85°C, amounts to a difference of 30°C. One of the common mistakes by beginners, using SOLIDWORKS and other finite element software, is to unwittingly consider the ambient temperature to be 0°C. If we had done that, a temperature difference of 54.85°C would have resulted.

After coupling the thermal effect with the static study, we are primed to deal with the other items in the static study.

Applying materials, fixtures, pressure, and runs

This section assumes you have already gone through the past chapters. This means you should be already familiar with how to assign material property, apply fixtures, apply external loads, and create a mesh and run the analysis. For this reason, the condensed steps to deal with the aforementioned items are summarized as follows:

1. For the material, assign **Chrome Stainless Steel**.

2. For the boundary application, apply a **Fixed Geometry** fixture, as shown in (a) in *Figure 9.16*.

3. For the load, apply an external pressure of **0.5** MPa to the top of the diaphragm, as shown in (b) in *Figure 9.16*:

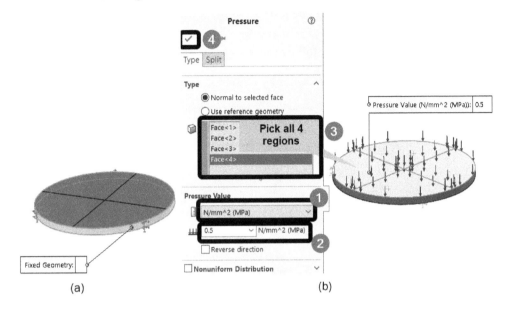

Figure 9.16 – Applications of fixed support and external pressure

4. Discretize the structure with a fine mesh via the curvature-based meshing algorithm, as shown in *Figure 9.17a*. The final meshed structure is depicted in *Figure 9.17b*:

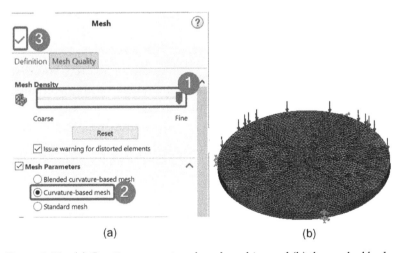

Figure 9.17 – (a) Creating a curvature-based meshing and (b) the meshed body

We are now set to run the analysis and obtain the results.

Running and examination of results

Our goal here is to obtain and compare the results of the static analysis with and without the thermal effect. For this purpose, follow these steps:

1. Run the analysis in the usual manner by following *Figure 9.18*:

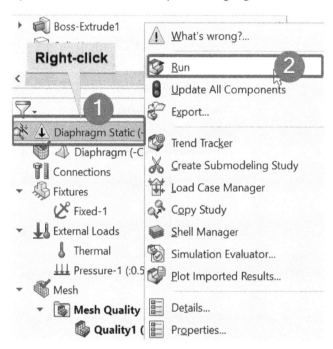

Figure 9.18 – Running the static study with the linked thermal load

With the running completed, the three default results of **Stress1(-vonMises-)**, **Displacement1 (-Res Disp-)**, and **Strain1(-Equivalent-)** will appear under the Results folder within the simulation study tree. For design purposes, it is good to track the resultant displacement and the von Mises stress results, both of which are shown in *Figure 9.19* and *Figure 9.20*:

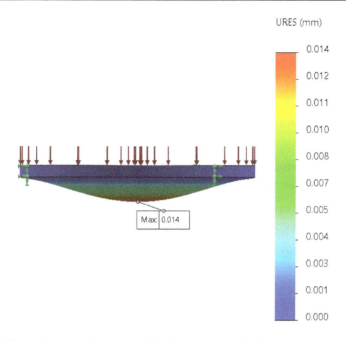

Figure 9.19 – The resultant displacement of the diaphragm under thermal and mechanical loads

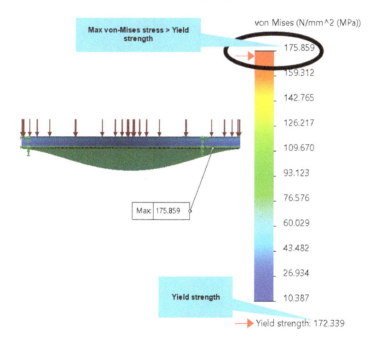

Figure 9.20 – The distribution of the von Mises stress of the diaphragm under thermal and mechanical loads

The preceding results amount to the combined effect of mechanical and thermal loads. *Figure 9.19* shows that the maximum resultant displacement is *0.014 mm*. Meanwhile, in *Figure 9.20*, you will notice that the von Mises stress surpassed the yield strength of the material, which means the combined effect of the thermal and mechanical loads generates huge stress that will cause the diaphragm to yield. In the subsequent subsection, we will explore how to steer the design away from failure.

Now, suppose that the diaphragm is subjected to the effect of *only the pressure load without the thermal effect* – what would be the magnitude of the stress in the diaphragm? To answer this question, we will carry out the next step.

2.   Right-click on the analysis name, and then select **Properties** (as we did in *Figure 9.5a*).

3.   When the static study dialog window appears, navigate to **Flow/Thermal Effects**, as shown in *Figure 9.21*, and then select the **Input temperature** box. This will deactivate the temperature load from the thermal study:

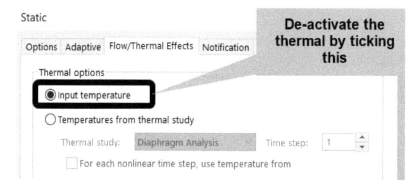

Figure 9.21 – A partial view of the static study dialog window to uncouple the thermal effect

4.   Click **OK** to close the Static Study PropertyManager.

As soon as *step 4* is completed, a warning symbol will appear beside the analysis name, as shown in *Figure 9.22a*, and the thermal load will disappear, as shown in *Figure 9.22b*:

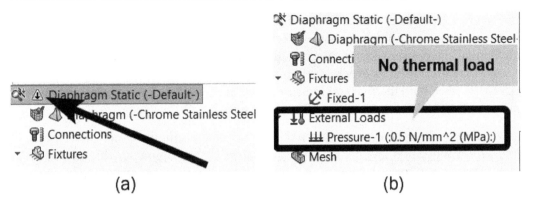

Figure 9.22 – (a) A warning to rerun the analysis and (b) the absence of a thermal load

5. Rerun the analysis.

From the results generated, *Figure 9.23 (a and b)* highlights the displacement and the von Mises stress of the diaphragm due to the pressure load effect alone:

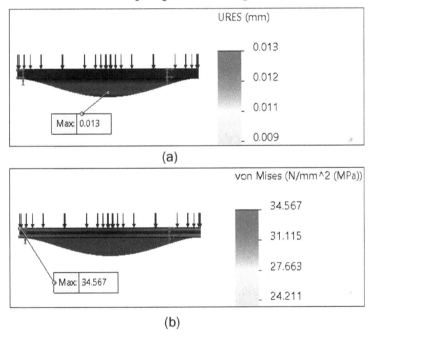

Figure 9.23 – Resultant displacement and the von Mises stress due to the pressure effect alone

From *Figure 9.23a*, we can see that the vertical displacement of the diaphragm is *0.013 mm* when we consider only the pressure effect. This is about *7.7 %* less compared to when both thermal and pressure effects were considered. Now, unlike the displacement value, the von Mises stress exhibits experience a much lower value of *34.56 MPa* compared to *184.39 MPa* when both thermal and pressure effects are considered, which amounts to a difference of *400%*! What is the cause of this huge stress?

You will observe that when we conducted the thermal analysis (in the *Dealing with the thermal study* subsection), we did not apply any fixture to the diaphragm. This means the structure was able to freely expand in all directions. By allowing it to expand freely, no stress will be developed within it. Thus, we have a case of *strain without stress*. However, within the static study environment, we applied pressure, considered the thermal effect, and constrained the diaphragm by preventing the movement of its edge. By constraining the diaphragm, the normal expansion that would have been caused by the rise in temperature is prevented. This then generates an additional compressive load to be imposed on the diaphragm in addition to the pressure load. So, in short, the higher stress during the combined thermal and static analysis boils down to the interaction between the pressure-induced stress and the internal compressive stress within the diaphragm resisting the temperature-induced expansion.

Recall that while reviewing *Figure 9.20* earlier, we mentioned that the huge von Mises stress under the combined thermal plus static study is more than the yield strength of chrome stainless steel, which signals failure. The crucial question we need to answer is how to achieve the goal of designing the diaphragm to be able to handle both the thermal and pressure loads without failing. For this, we shall introduce the concept of optimization study in the next section.

Updating the geometry and rerunning the analysis

Assuming the material selection option is restricted to chrome stainless steel, and we now wish to determine the combination of diameter and thickness that will produce an acceptable stress level within the diaphragm, we may either resort to a crude trial-and-error method (which is akin to searching in the dark while being blindfolded) or adopt a more systematic optimization. To showcase another important capability of SOLIDWORKS Simulation, we will embrace the optimization approach.

We will lay out the procedure for the optimization in the next few pages, so follow along to complete to steps:

1. Include the thermal effect in the static analysis setup and rerun the analysis (see *Figure 9.14*).

2. At the bottom of the graphics window, right-click the static study tab and select **Create New Design Study**, as shown in *Figure 9.24*:

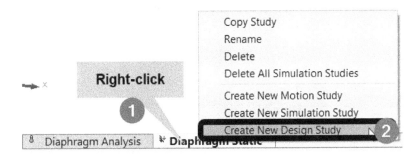

Figure 9.24 – Initiating the New Design Study tab

The preceding step will launch the optimization design environment with the default look, as shown in *Figure 9.25*. As you can see from this figure, we need to specify three sets of optimization parameters in the form of **Variables**, **Constraints**, and **Goals**:

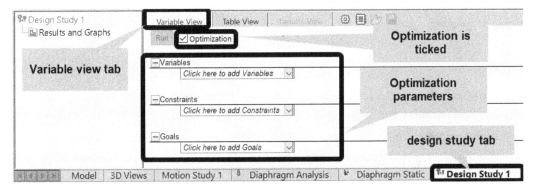

Figure 9.25 – Optimization study

Let's briefly take a look at what each of these means:

A. *Variables*: This refers to factors within the design that are allowed to change. For many simulation tasks, this often involves geometric features of our design.

B. *Constraints*: This represents a condition that our final design should satisfy. For most static analyses, this may involve reducing the deformation, strain, or stress within our design to an acceptable level.

C. *Goals*: This generally refers to something we wish to minimize or maximize.

We will specify these three parameters in the upcoming steps.

3. Under **Variables**, click on the drop-down menu arrow, and then click on **Add Parameter**, as shown in *Figure 9.26a*. This action will launch the **Parameters** manager that is partially shown in *Figure 9.26b*:

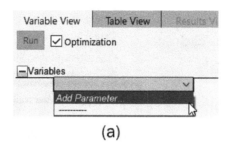

(a)

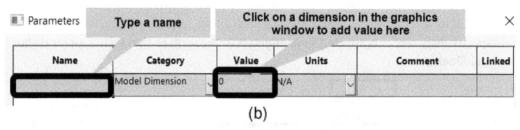

(b)

Figure 9.26 – Adding a parameter under the optimization variable

Within the **Parameters** window, our focus will be on the first three columns – **Name, Category**, and **Value**.

4. In the column named **Category**, the **Model Dimension** option is selected by default, as shown in *Figure 9.26b*; keep it as such.

5. Now, click inside the box under **Value**, and then navigate to the graphics window to click on the diameter dimension, as shown in *Figure 9.27*. This will instantly fill the box with the value of 60, which is the diameter of the diaphragm.

6. Click **OK**:

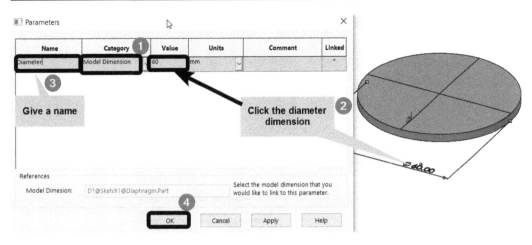

Figure 9.27 – Adding the diaphragm's diameter as one of the optimization variables

7. Repeat *steps 2–5* to add thickness as another parameter, as shown in *Figure 9.28*, and then click **OK** to wrap up the selection:

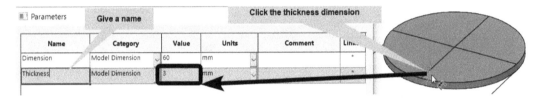

Figure 9.28 – Adding the diaphragm's thickness as the second optimization variable

8. With *step 6* completed, you will be back at the **Variable View** tab. From here, change the interval boxes for each variable to the **Range** type and set the upper and lower limits, as shown in *Figure 9.29*:

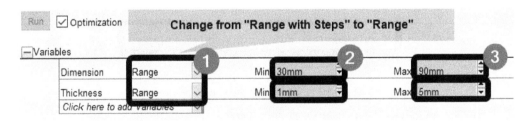

Figure 9.29 – Updating the variables interval type and ranges

9. Still within the **Variable View** tab, under **Constraints**, click on the drop-down menu arrow, and then click on **Add Sensor**, as shown in *Figure 9.30a*. This action will launch the constraint's **Sensor** property manager.

10. Within the **Sensor** property manager, set the options shown in *Figure 9.30b*:

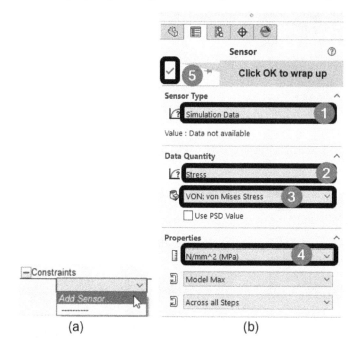

(a)   (b)

Figure 9.30 – Adding and updating the sensor details for the constraint

By completing *step 9*, you will be back at the **Variable View** tab again; make the following changes.

11. Under **Constraints**, change the sensor type to **Is less than** and set the **Max** value to **145** MPa, as indicated in *Figure 9.31*:

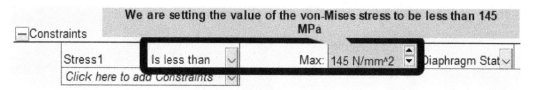

Figure 9.31 – Setting the limit of the von-Mises stress

12. Under **Goals**, click on the drop-down menu arrow, and then click on **Add Sensor**, as shown in *Figure 9.32a*. This action will launch the goal's **Sensor** property manager.

13. Within the goal's **Sensor** property manager, set the options shown in *Figure 9.32b*:

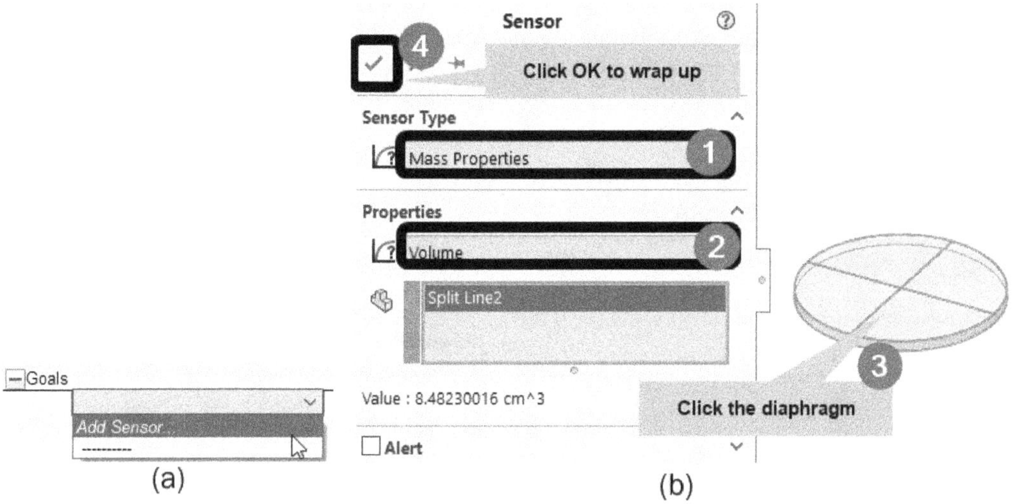

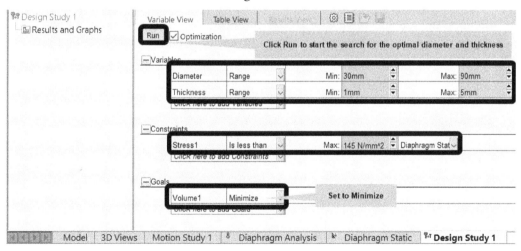

Figure 9.32 – Adding and updating the sensor parameter

14. By completing *step 12*, you will be returned to the **Variable View** tab; ensure that the goal option is set to **Minimize**.

    Once all the three important parameters are set, the **Run** button will become active, and the interface should look like *Figure 9.33*:

Figure 9.33 – The complete setup of the optimization study

> **A Summary of the Optimization Setup**
>
> Find the combination of diameter and thickness within the specified range specified that will have a minimum volume of material to be used to design the diaphragm, while satisfying the condition of having a von Mises stress that is less than 145 MPa.

15. Click **Run** to begin the optimization search, which will launch the optimization iteration window shown in *Figure 9.34*:

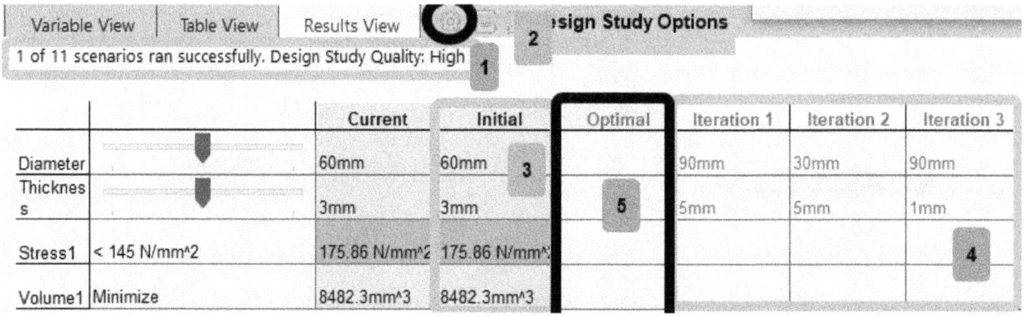

Figure 9.34 – Results View with the iteration window

Let's briefly discuss a few items from *Figure 9.34*:

- The first is the fact that the optimization iteration involves 11 scenarios, as indicated in the item labeled 1. Still with the item labeled 1, notice that the status of the study option is indicated by the phrase **Design Study Quality: High**. Typically, **High** is the default option, but you can set the quality by using the **Design Study Options** button labeled 2.

- Next, the initial values of the design parameters are indicated in the box labeled 3, while the combinations of various values of diameter and thickness are contained in the partially shown iterations table labeled 4.

At the end of the optimization run, the optimization algorithm will sort through the iteration to find the set of parameters that fits our design constraint (that is, with a von Mises stress less than 145 MPa) while meeting the goal of minimizing the volume. Once found, the column named **Optimal** will be filled accordingly. If no optimal solution is found, then you will get an **Optimization failed** message.

Fortunately, the running of the current study is a success, as shown in *Figure 9.35*:

		Current	Initial	Optimal	Iteration 1	Iteration 2
Diameter		30.0123mm	60mm	30.0123mm	90mm	30mm
Thickness		1.68332mm	3mm	1.68332mm	5mm	5mm
Stress1	< 145 N/mm^2	137.71 N/mm^2	175.86 N/mm^2	137.71 N/mm^2	281.7 N/mm^2	203.28 N/mm^2
Volume1	Minimize	1190.85mm^3	8482.3mm^3	1190.85mm^3	31808.6mm^3	3534.29mm^3

Figure 9.35 – Results View with the optimal solution

As you can see from *Figure 9.35*, within the range of values that we have considered, the optimization result yields an optimal solution that turns out to be a combination of diameter and thickness values of *30.01 mm* and *1.68 mm* respectively. How does this reduction impact the performance of the diaphragm? For one, this combination yields a von Mises stress of *137.71 MPa*, which is 27% lower than the value of *175.86 MPa* we started with. Additionally, the von Mises stress we obtained is also very much lower than the yield strength of chrome stainless steel, which means the design is now much safer. Relatedly, you will also see that the new volume of *1190.85 mm³* is lower than the initial volume of *8482 mm³* (see the **Optimal** and **Initial** columns in *Figure 9.35*).

Finally, by clicking on any of the iteration columns, you would make its result active in the graphics environment and the associated study environments. For instance, the **Optimal** column has been clicked to make it the current study. After making it the current study, you can then switch to the coupled static study environment. Upon switching, you will spot two artifacts of the optimization study, as illustrated by the von Mises stress plot shown in *Figure 9.36*. The first artifact to note is the presence of the two sensor items (which we created in *steps 9* and *12*) under the features manager tree (labeled 1). Second, there will be a new item named **Parameters** (labeled 2) embedded within the simulation study tree:

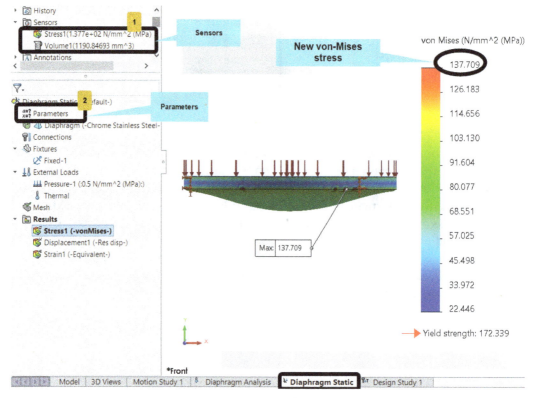

Figure 9.36 – Examining the updated von Mises stress based on the optimal parameters

You can right-click on the two artifacts highlighted previously for further explorations, and you can examine other results obtained in the case of combined static and thermal study. Meanwhile, for verification purposes, three more results are shown in *Figure 9.37*, *Figure 9.38*, and *Figure 9.39*:

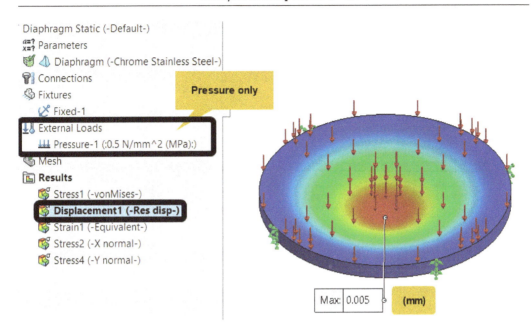

Figure 9.37 – The resultant displacement with the optimal parameters under pressure load alone

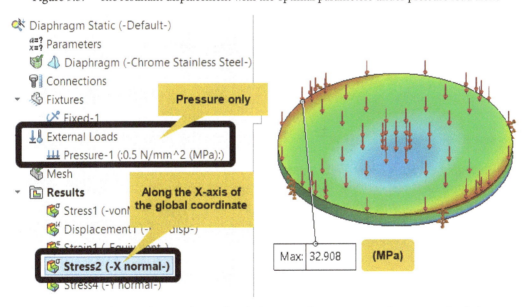

Figure 9.38 – The radial stress obtained with the optimal parameters under pressure load alone

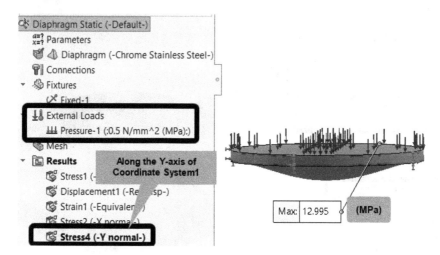

Figure 9.39 – The tangential stress obtained with the optimal parameters under pressure load alone

For the three preceding results, the thermal analysis was deactivated (as done earlier in *Figure 9.21*) so that only pressure load is used for the analysis (using post-optimization geometric data). Interestingly, these last three results can be compared with the output from the following expressions for the axisymmetric analysis of thin plates [4]:

- y_{max} (occurs at the center of the plate) $= \dfrac{3(1 - v^2)\, pa^4}{16Et^3}$ (9 - 1)

- Maximum radial stress $= \pm\dfrac{3}{4}\dfrac{pa^2}{t^2}$ (9 - 2)

- Maximum tangential stress $= \pm\dfrac{3}{8}\dfrac{pa^2}{t^2}(1 + v)$ (9 - 3)

In the preceding equations, v and E are material properties denoting the Poisson's ratio and Young's modulus respectively. The a and t parameters symbolize the radius and thickness of the diaphragm, while p is the applied pressure. By substituting the value of $v = 0.28$, $p = 0.5$ MPa, $E = 200 \times 10^9$, $t = 1.68$ mm, and $a = d/2 = 15$ mm in the preceding theoretical equations, we obtain the value of *0.0046 mm* for the maximum deflection, *29.99 MPa* for the radial stress, and *14.95 MPa* for the tangential stress. As you can see, the theoretical deflection value differs from the one computed by SOLIDWORKS by just *8.7%*, while the stresses differ by *10%* and *13%* respectively. Take note that the high difference between the values from the theoretical expressions and the simulation results can be attributed to the fact that a collection of solid elements is used for the study (for thin plates, shell elements, covered in *Chapter 5, Analyses of Axisymmetric Bodies*, give better accuracy).

This concludes the solution to the case study on the exploratory design analysis of a diaphragm under the combined influence of temperature and mechanical loads. Over the last several pages, we have described the procedure for coupling a thermal study with a static study. Along the way, we conveyed the strategy for including and excluding temperature effects in a static study and demonstrated how to employ SOLIDWORKS Simulation's optimization capability to obtain optimal geometric parameters for the problem studied. Moving forward, it is necessary to emphasize that apart from thermal loads, another type of load that is of importance to design engineers is cyclic load, which is the focus of the next section.

Analysis of components under cyclic loads

The effect of cyclic loads is closely related to fatigue failure, which is another broad topic on its own. In the past chapters, we have based the failure assessments of components that we studied on the idea that failure will happen if a specific stress measure exceeds the yield strength (for ductile components) or the ultimate strength (for brittle components). With fatigue failure, the stress required to bring a component to failure is often far lesser than the yield or ultimate strengths of the material under study. According to numerous studies, more than 50% of machinery breakage can be attributed to fatigue failure [4, 5]. While this section is not intended to cover fatigue failure in detail, we will outline four major concepts that you need to be aware of to conduct a basic fatigue analysis:

- *Stress-time or load-time cycle*: This is the plot of stress/load versus time that describes the cyclical nature of the stress/load applied to a component. Different forms of this curve occur in a variety of practical scenarios. Two common examples shown in *Figure 9.40* are as follows:

  A. The fully reversed stress cycle, which is most commonly used for laboratory fatigue tests, and where the maximum and minimum stresses are equal in magnitude but opposite in sign.

B.   The zero-based stress cycle where the minimum stress is zero but the maximum stress is some other positive number:

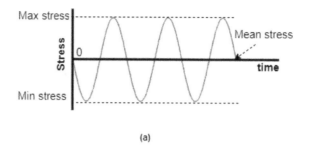

(a)

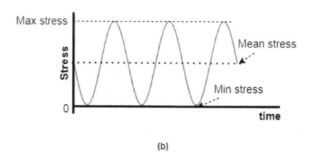

(b)

Figure 9.40 – Two examples of stress cycles

- *S-N or the Wöhler curve*: This is the curve of stress versus the number of cycles to fracture (that is, when the component breaks apart). Most S-N curves are based on the fully reversed load/stress cycles. For cases of loading in which the stress/load cycle is not fully reversed, empirical fatigue expressions such as **Goodman**, **Soderberg,** and **Gerber** relations must be used to complement the S-N curve.

- *Fatigue limit/endurance limit*: This refers to the stress level below which a component can undergo an infinite number of stress cycles without failure. It is an important number that is determined by an S-N curve.

- *Fatigue life*: This is an important outcome of interest in fatigue analysis. It refers to the number of cycles that a component is subjected to under a specific load before fracture takes place. Quite often, fatigue life is affected by many factors, including environmental factors, microstructural flaws in materials, manufacturing-induced defects, the presence of stress raisers, and so on *[6]*.

With this admittedly brief background, let's now examine an example that features these concepts in the context of SOLIDWORKS Simulation.

Problem statement

A thin stepped bar of 10 mm thickness is subjected to a static pressure load of magnitude *200 MPa*, as shown in *Figure 9.41*. The bar is made of alloy steel with a yield strength of *620 MPa* and an ultimate strength of *810 MPa*. Our objectives with this problem is as follows:

- To conduct a static analysis to determine the stress in the component with a non-cyclical application of the load

- To connect the previous static study with a fatigue analysis of the component under a fully reversed cyclical application of the pressure load to determine its fatigue life and to verify whether or not fatigue failure occurs in the component based on a design life of 10^6 cycles:

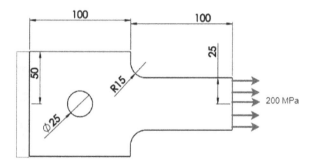

Figure 9.41 – A tensioned steel bar with two stress raisers

The next section addresses the problem.

The solution to fatigue analysis

As with the thermo-mechanical analysis conducted in the previous section, a preliminary static analysis is generally conducted prior to doing a fatigue analysis. To make the presentation concise, a file is presented that contains an implemented static analysis, which is what is reviewed next.

Review of the static study

To begin this exercise, download the file named `FilletedBar` within the `Chapter 9` folder that you downloaded for the previous exercise. Given that you are now fully familiar with the setup for static studies, the file contains both the geometric model and a completed static study:

Let's now review the file:

1. Open the part file (`FilletedBar`) via **File → Open**.

2. If the simulation add tab has not been initiated, then activate it via the **Command manager** tab done as we have been doing.

3. Once the simulation tab is active, you will see the static study tab; rerun the analysis and the simulation items will appear, as shown in *Figure 9.42*:

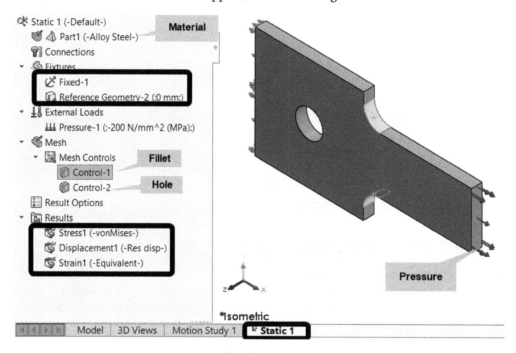

Figure 9.42 – Examining the implemented static study

4. Double-click on the **Stress1** result to visualize the von Mises stress (*Figure 9.43*):

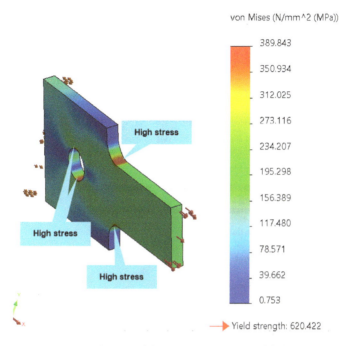

Figure 9.43 – Distribution of the von Mises stress under the static load

Note that the maximum von Mises stress (*389.934 MPa*) revealed in *Figure 9.43* is less than the yield strength of the material (*620 MPa*). This gives the impression that the component is safe. In the next subsection, we will strive to draw further insight into the safety of the component from a fatigue analysis.

Fatigue analysis

Here, we will check for the possibility of fatigue failure if the component is subjected to a cyclical form of the preceding stress. The procedure for fatigue study comprises the following steps:

1. Navigate to the base of the graphics window, right-click on the study tab, and then select **New Simulation Study** (as we did in *Figure 9.12a*).

2. Within the **Study** property manager that appears, under **Advanced Simulation**, click **Fatigue**.

3. Select the **Constant amplitude events with defined cycles** option (labeled 2 in *Figure 9.44*).

4. Provide a name for the new study and wrap up by clicking **OK**, as shown in *Figure 9.44*:

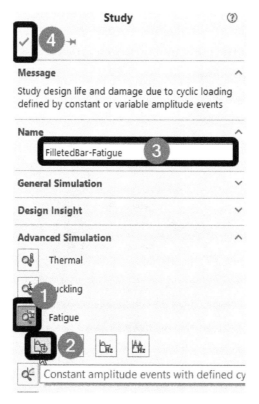

Figure 9.44 – Initiating the fatigue study environment

5. Within the fatigue study simulation tree items, right-click the study name and choose **Properties**, as illustrated in *Figure 9.45*:

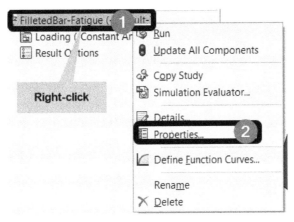

Figure 9.45 – Modifying the fatigue study property

This launches the fatigue study dialog box, within which we need to make the following options.

6. Within the **Options** tab, make the selections shown in *Figure 9.46*:

Figure 9.46 – Fatigue analysis dialog box

Once you conclude the step by clicking **OK**, the fatigue study environment will appear. But before we move onto the environment, a couple of comments are desired about the selections in *Figure 9.46*.

Note that we have four options under the **Mean stress correction** item (labeled 3). The first option is most appropriate for a fully reversed stress cycle (such as *Figure 9.40a*). For a fully reversed stress cycle, the mean stress is zero, and hence there is no need for a mean stress correction factor. However, the other three options, which we also mentioned in the introduction to this section, are used for non-fully reversed stress cycles (such as *Figure 9.40b*). Among these other options, the most conservative and thus preferable for non-fully reversed stress cycles is the **Goodman** correction factor. Next, in the box labeled 5, we have chosen the upper limit of 1,000,000 cycles for the fatigue life (N). This number is used because the fatigue strength of most materials is often is specified at $N > 10^6$. Furthermore, close to the box labeled 5, note that we maintained the value of 1 for the fatigue reduction factor. For other practical situations, a value less than 1 but greater than 0 may be used to quantify the reduction in the fatigue strength of the component due to the presence of the fillets for instance. Let's now complete the other simulation tree items.

7. Under the fatigue simulation study tree, right-click **Loading** and choose **Add Event** as shown in *Figure 9.47a*.

8. Within the **Add Event** property manager that appears, follow the screenshot shown in *Figure 9.47b*:

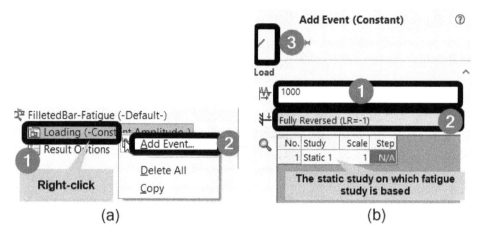

Figure 9.47 – Specifying the fatigue load parameters

By completing *step 8*, a new item will appear within the simulation study tree bearing the name of the model – in this case, FilletedBar.

9. Right-click on the FilletedBar item that appears, and then click **Apply/Edit Fatigue Data** (*Figure 9.48*):

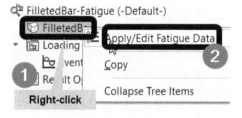

Figure 9.48 – Initiating the application of material property

10. Within the material database, follow the selections depicted in *Figure 9.49*:

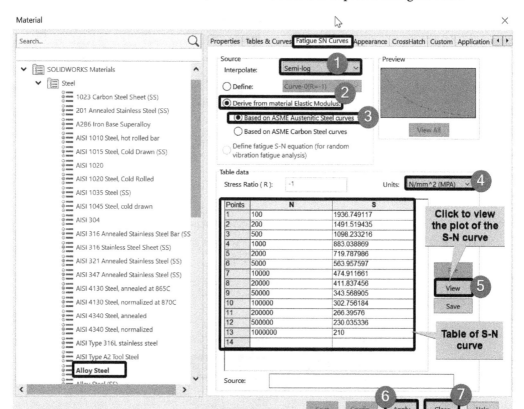

Figure 9.49 – Material database

At the center of *Figure 9.49* is the table of the S-N curve, which was among the concepts discussed in the introduction to this section. A specific row of the N column represents the number of cycles at which the corresponding stress value in the S column can be applied before fatigue failure. As you can see, the last row of the table indicates that $S = 210\ MPa$ is the stress that corresponds to N = 10^6. This means that the fatigue strength of this material is *210 MPa*. So, all things being equal, components derived from this material can be subjected to an infinite cycle of load at stress values that equal or fall below *210 MPa* without fatigue failure. There's one more thing – by default, the plot of the S-N plot shown in the top right corner is displayed as a log-log plot. However, it is generally easier to work out the stress that corresponds to the infinite life of the material by changing to the **Semi-log** option, as done in the box labeled 1. For a better view of the S-N curve, you can click on the button labeled 5, which will create a new window of the S-N curve, through which you can observe that the curve becomes horizontal at $S = 210\ MPa$.

At this stage, we are done with the setup for fatigue analysis, and we are ready to run the analysis.

11. Run the analysis, as shown in *Figure 9.50*:

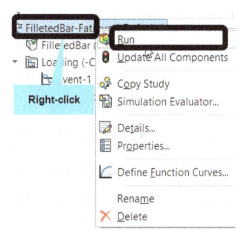

Figure 9.50 – The running of the fatigue study

Once the running is complete, you will notice **Results1 (-Damage-)** and **Results2 (-Life-)** under the Results folder. By double-clicking on each, you can reveal the plot of the distribution of the fatigue damage (*Figure 9.51*) and fatigue life (*Figure 9.52*):

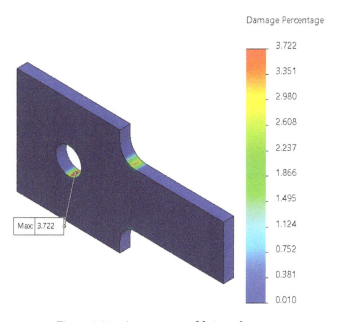

Figure 9.51 – Assessments of fatigue damage

Referring to *Figure 9.51* and *Figure 9.52*, it is obvious that the stress hot spots are located at the fillets and the top and bottom of the hole. This is the same area that the static result revealed in *Figure 9.43*. However, while the static result gave an impression of safety, *Figure 9.52* shows that the fatigue capacity of the component around the hole is significantly less, with a life of *26,828* for a load block of 1,000 (see *Figure 9.47b* for the load block). According to the literature on the theory of machine design *[6]*, this component will be categorized as having a finite fatigue life, given that it has a region with fatigue capacity in the $10^3 < N < 10^6$ range.

Finally, the trend of the results in *Figure 9.51* and *Figure 9.52* is consistent with well-established observations that fatigue failure is naturally initiated by discontinuities in the form of a rapid change in cross section (such as the filleted region and the hole located within the studied component) *[6]*. Under a repeated cyclical load, discontinuities represent fertile ground for the initiation and propagation of cracks that are closely linked with the fatigue failure mechanism. The good news is that the power of finite element simulation can always be leveraged to investigate the effect of discontinuity such as fillets, keyways, and grooves, as we've done here. However, other critical factors, such as environmental influences in the form of corrosion, surface defects, fabrication flaws, and microscopic defects/inclusions, cannot be easily captured without resorting to advanced fatigue/fracture simulation platforms such as NASGRO, developed by NASA *[7]*:

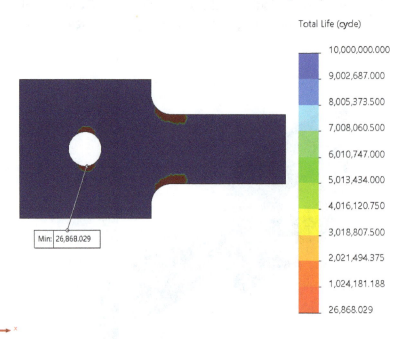

Figure 9.52 – A plot of the fatigue life

This ends our exploration of the solution to the problem posed at the beginning of this section. Overall, this last example serves only to give an introductory coverage of the fatigue analysis of a component under the effect of a cyclical load using SOLIDWORKS Simulation. Understandably, we have only looked at a constant amplitude stress cycle premised on a static load. However, SOLIDWORKS Simulation is also capable of dealing with variable amplitude stress cycles premised on a static load. Moreover, it is capable of handling constant/variable amplitude stress cycles based on dynamic vibration loads and analyses, based on more than one static or dynamic study. In all, fatigue analysis is a very delicate and daunting task for most design engineers. Nevertheless, it is hoped that this introduction guides you toward taking a journey to discover other advanced features of this aspect of the simulation environment for more complex analysis.

Summary

This chapter covered the procedures needed to factor in thermal and repeated load cycle effects in the analysis of engineering components, using two hands-on examples. In addressing the problems framed around the examples, we showcased the following strategies:

- How to integrate thermal and static analyses to address the simulation of components at an elevated temperature

- How to investigate the fatigue life for components under the effect of cyclical loads

- How to reap the benefit of the optimization capability of SOLIDWORKS Simulation to design components against failure

These concepts add to your repertoire of analysis techniques, which can be leveraged for a comprehensive assessment of a wide variety of design problems that you are likely to encounter.

In the next chapter, we will round up our coverage of the topics by re-examining certain aspects of meshing that will further solidify your exposure to finite element simulation.

Further reading

- [1] *Pressure sensors: The design engineer's guide, AVNET* https://www.avnet.com/wps/portal/abacus/solutions/technologies/sensors/pressure-sensors/ (accessed 20/08/2021)

- [2] *Design for Thermal Stresses, R. F. Barron and B. R. Barron, Wiley, 2011*

- [3] *Thermal Analysis with SOLIDWORKS Simulation 2019 and Flow Simulation 2019, P. Kurowski, SDC Publications, 2019*

- [4] *Advanced Mechanics of Materials, A. P. Boresi, R. J. Schmidt, and Knovel, Wiley, 2003*

- [5] *Advanced Mechanics of Materials, R. D. Cook and W. C. Young, Prentice Hall, 1999*

- [6] *Shigley's Mechanical Engineering Design, R. Budynas and K. Nisbett, McGraw Hill Education, 2010*

- [7] NASGRO, *NASA/Southwest Research Institute* https://www.swri.org/nasgro-software-overview (accessed 20/08/2021)

10

A Guide to Meshing in SOLIDWORKS

Meshing is the process of discretizing a structure into smaller, finite substructures. As you will have observed from past chapters, a strong interplay exists between meshing and results obtained from finite element analysis. For this reason, we will use this chapter to complete our exploration of finite element simulation for static analysis by taking a careful look at ways to customize the meshing of a structure to achieve reliable results. You will encounter mesh control (again) and learn how to employ convergence analysis to evaluate the accuracy of simulation results. In pursuit of these ideas, the remainder of the chapter entails discussion centered around the following topics:

- Discretization with beam and truss elements
- Mesh control with plane elements
- Mesh control with three-dimensional elements
- Discretization with h- and p-elements

Technical requirements

You will need to have access to the SOLIDWORKS software with a SOLIDWORKS Simulation license.

You can find the folder containing the models required for this chapter here:

```
https://github.com/PacktPublishing/Practical-Finite-Element-
Simulations-with-SOLIDWORKS-2022/tree/main/Chapter10
```

Discretization with beam and truss elements

We covered the discretization of structures with a collection of truss elements in *Chapter 2, Analyses of Bars and Trusses*, and with a collection of beam elements in *Chapter 3, Analyses of Beams and Frames*, and *Chapter 4, Analyses of Torsionally Loaded Components*. Let's briefly highlight two of the ideas covered in those past chapters:

- We discussed the characteristics of these elements in terms of degrees of freedom – specifically, that SOLIDWORKS truss element has 3 degrees of freedom per node, while the beam element has 6 degrees of freedom per node.

- We outlined a few tricks to be employed when using those elements. Specifically, for beam elements, we discussed the idea of taking advantage of critical positions under the *Modeling strategy* subsection in *Chapter 3, Analyses of Beams and Frames*.

In addition to the preceding points, there is one important distinction between these two elements that you should know. SOLIDWORKS Simulation will use multiple beam elements to discretize a single beam-like structure. However, it uses a single truss element to discretize a truss-like structure. The implication of this is that you can specify the number of beam elements to be used in a simulation through mesh control, but you cannot use mesh control for truss elements. Recall that we introduced mesh control for beams in *Chapter 7, Analyses of Components with Mixed Elements*, in the *Part C – Meshing and running* section.

To get to the heart of the preceding point, consider the problem of analyzing the suspender system displayed in *Figure 10.1*, which is inspired by *Exercise 4.55* in the book by *Hibbeler* [1]. The system comprises three links and a beam that is used to support two loads, as shown in the following figure. It is necessary to determine the normal stress of the beam:

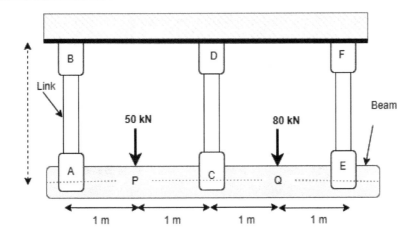

Figure 10.1 – A three-bar suspender system

In solving the problem, *the links AB, DC and FE* are discretized with truss elements. Next, by employing the idea of critical positions, which we covered in *Chapter 3, Analyses of Beams and Frames*, we would have four beam segments (*AP, PC, CQ and QE*) during the modeling phase. Finally, each of the beam segments will be divided into smaller beam elements as part of the discretization process.

Let's now look at a complete solution to the preceding problem, which is included in the Chapter 10 folder that you can download from this book's GitHub repository. After downloading the folder, follow these steps to examine the simulation studies:

1. Open the SOLIDWORKS part file named ThreeBars_Beam.
2. Activate SOLIDWORKS Simulation. Once the simulation add-in is activated, you will see the two studies, as shown in *Figure 10.2*:

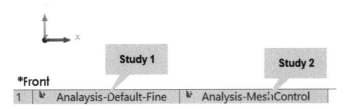

Figure 10.2 – Static studies contained within the part file

3. Review the simulation study tree items shown in *Figure 10.3*. You will notice that they comprise a combination of truss and beams used for the discretization:

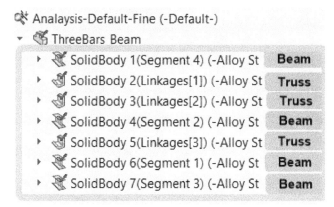

Figure 10.3 – Discretizing with beam and truss elements

4. Review the mesh details, as shown in *Figure 10.4*:

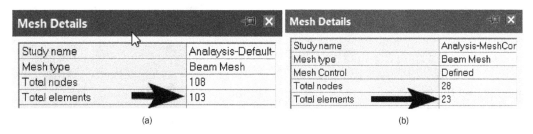

Figure 10.4 – (a) Study 1 mesh and (b) study 2 mesh details

5. Examine the results of bending stress from both studies, as shown in *Figure 10.5*:

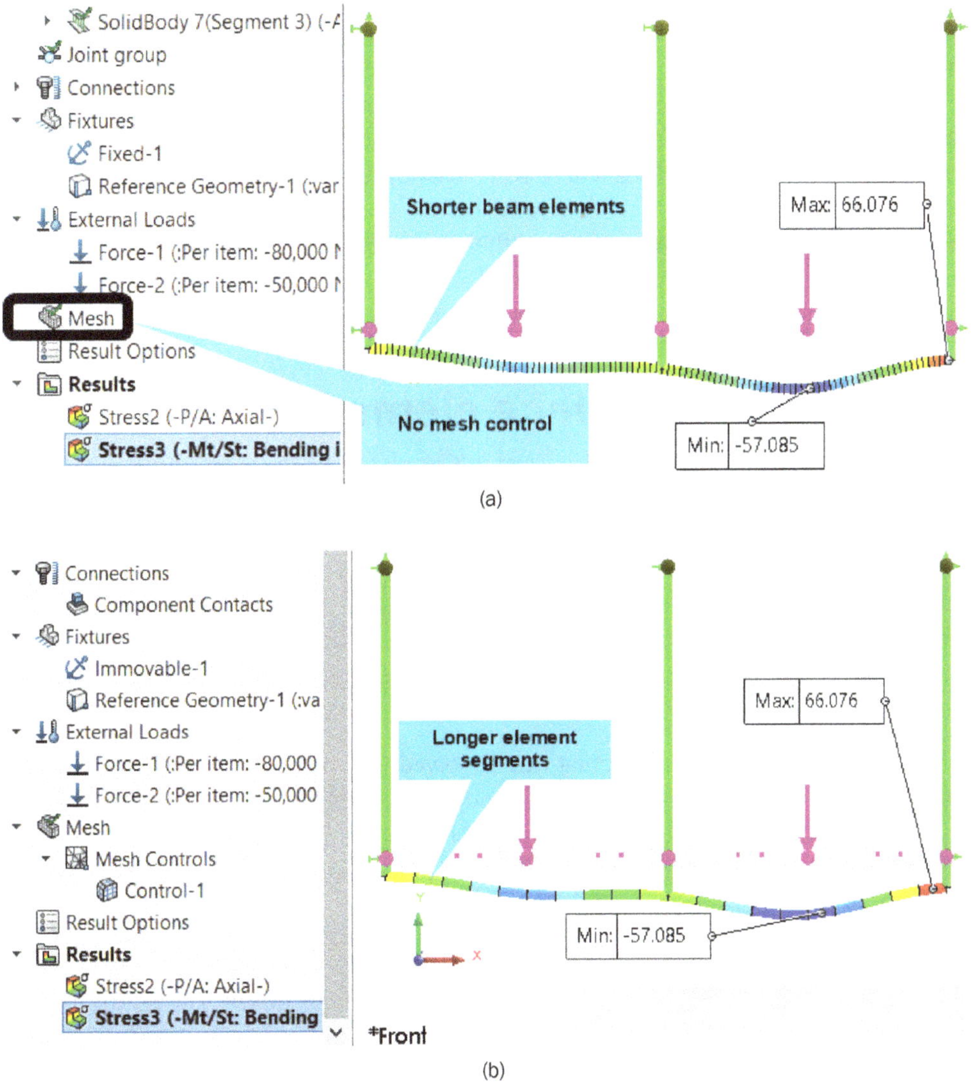

(a)

(b)

Figure 10.5 – (a) Bending stress from study 1 and (b) bending stress from study 2

What can we learn from the presented results? First, as you can see from *Figure 10.4*, study 1 has a total of 103 elements because we used a fine mesh, resulting in 100 shorter beam elements, plus 3 truss elements. In contrast, we utilized a moderate mesh for study 2, which yields 23 elements (20 beam elements, plus 3 truss elements). You will get 20 if you count the beam segments in *Figure 10.5b*.

Next, as you can see by comparing the results of the two studies depicted in *Figure 10.5*, the values of the maximum and minimum bending stresses from both studies are virtually the same, even though we used a reduced number of elements for study 2. Generally, it is often tempting to use a fine mesh, as done in study 1, but for large-scale structures, doing so can quickly overwhelm your system's memory. Therefore, by exploiting mesh control to customize the number of beam elements for your study, you will be able to shorten the simulation runtime in situations where you have hundreds or thousands of beam structures forming a component, as is often the case in the automotive and aerospace industries. Now, you may be wondering how then do we reconcile the number of elements to use in mesh control with the accuracy of the simulation results? The answer is premised on what is called mesh convergence, which is featured in the next section.

Mesh control with plane elements

Plane elements were introduced and discussed in *Chapter 5, Analyses of Axisymmetric Bodies*, where it was emphasized that they represent a two-dimensional approximation of three-dimensional continuum structures. As you will recall from that chapter, we showcased the application of the element under the *Plane analysis of axisymmetric bodies* subsection. Here, we will highlight the benefit of using mesh control with plane elements for an analysis that centers around plane stress. As you will know, a plane stress problem involves *"a state of stress in which the normal and shear stresses directed perpendicular to the plane are assumed to be zero"* [2].

Figure 10.6 is indicative of a problem that can be solved with a collection of plane elements. We shall use the determination of the maximum normal stress that develops in the component upon loading to demonstrate the importance of mesh control and convergence study:

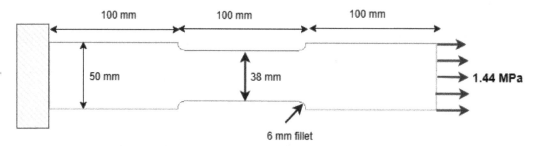

Figure 10.6 – A filleted specimen under a plane stress loading condition

To support our exploration of the concept of mesh control and convergence study, a complete solution is included in the file named `PlaneStress_Bar` within the `Chapter 10` folder.

As we did earlier, follow these steps to examine the simulation studies:

1. Open the SOLIDWORKS part file (`PlaneStress_Bar`).

2. Activate SOLIDWORKS Simulation (if you are starting SOLIDWORKS afresh).

    Examine the studies, as shown in *Figure 10.7*. As you can see, there are seven studies within the file:

Figure 10.7 – Seven simulation studies for the plane stress problem

3. Review the Simulation study tree items for all studies.

4. Next, review the mesh setting for each study (simply right-click the **Mesh** property in the Simulation study tree and choose **Create Mesh**).

    As you switch from one tab to another to examine the studies, you will notice that the only difference between the seven studies is the mesh quality. All other settings involving material, fixtures, and external loads are the same as presented in *Figure 10.8*:

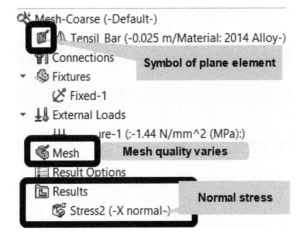

Figure 10.8 – Simulation study tree items with plane elements

5. After examining the mesh setting, you should click **Cancel** to close the **Mesh** property manager.

As you review the mesh setting, you will observe that we first start with a coarse mesh for the first study (*Figure 10.9a*), then a moderate mesh in the second study (*Figure 10.9b*), and then a fine mesh in the third study (*Figure 10.9c*). Take note of the position of the mesh slider and the element size (approximately 9.4 mm, 4.8 mm, and 2.4 mm respectively):

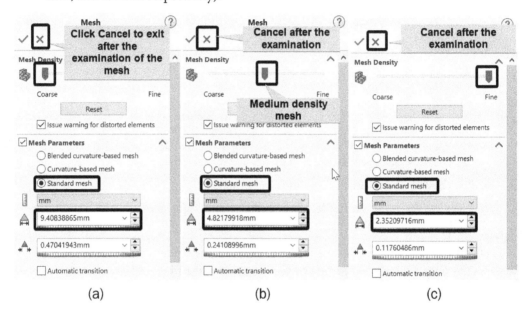

(a)   (b)   (c)

Figure 10.9 – Global mesh quality adjustment for the first three studies

For brevity's sake, the mesh details of the remaining four studies are not shown. However, they are based on the combination of the fine mesh in *Figure 10.9c* and mesh control features. To see the reason for the mesh control, let's review the results that correspond to the preceding three studies, as shown in *Figure 10.10*, depicting the distribution and maximum value of the axial normal stress:

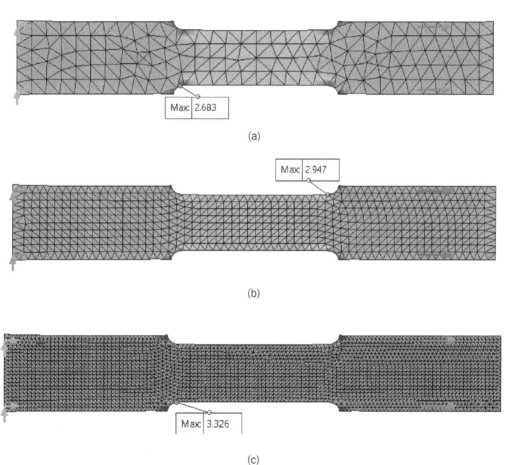

Figure 10.10 – Maximum principal stress (MPa) – (a) coarse mesh, (b) moderate mesh, and (c) fine mesh

An important insight from *Figure 10.10* is that, in each stress plot, some areas exhibit a uniform and smooth color transition (mostly away from the fillets). This indicates that the mesh size is adequate to get accurate results in those regions. However, as you can see, there is a mixture of color in the regions close to the four fillets. This means the stress value has not converged in those regions. Related to this, the plots show that the value of the maximum normal stress changes as we vary the mesh size from a coarse mesh (*2.683 MPa*) to a fine mesh (*3.326 MPa*). By using local mesh control for the filleted regions, we may be able to have a better estimate of the maximum stress around the fillets. This is what we have done in the remaining four studies shown in *Figure 10.7*! A representative demonstration of the mesh control carried out in the last study (labeled 7 in *Figure 10.7*) is portrayed in *Figure 10.11*:

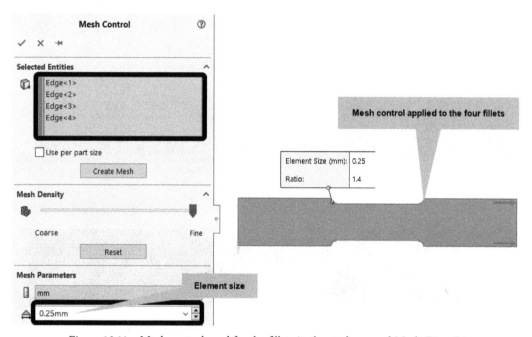

Figure 10.11 – Mesh control used for the fillets in the study named Mesh-Fine-C4

As you can see, we have specified an element size of 0.25 mm for the fillet for the aforementioned study. For the other three studies, you can check to confirm that the element size for the fillet is specified as 2 mm, 1 mm, and 0.5 mm respectively. By running the analysis for each of these mesh refinements, we can compile and tabulate the results, as shown in *Table 10.12*:

Study name	Number of Elements	Maximum normal stress (MPa)
Mesh-Coarse	312	2.683
Mesh-Moderate	1170	2.947
Mesh-Fine	4964	3.326
Mesh-Fine-C1	4976	3.339
Mesh-Fine-C2	5092	3.368
Mesh-Fine-C3	5450	3.427
Mesh-Fine-C4	6418	3.436

Table 10.12 – The variation of element size and normal stress for the seven studies

If we plot the number of elements against the maximum normal stress shown in *Figure 10.12*, then we can observe the trend of the curve flattening as the element number reaches close to 6,000:

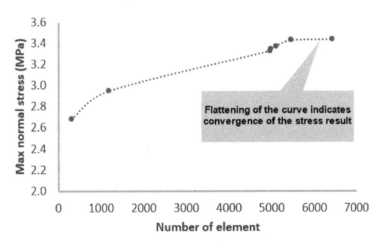

Figure 10.12 – A visualization of the convergence of results

The flattening of the curve suggests that the result is converging toward a finite value. Consequently, what we've done here is called a *convergence study*. It simply involves the setting up of simulation studies to observe how a particular simulation result converges toward a finite value. For static studies, a result of interest could be the maximum displacement, a specific reaction force stress value (such as principal stresses or the von-Mises stress). The plot can also be against the element size, the number of degrees of freedom, and so on. Furthermore, the mesh control can be as follows:

- Global – where the mesh is gradually refined in all areas of the structure
- Local – where the mesh refinement is focused on certain intricate features, such as holes, grooves, and fillets
- Hybrid – where the refinement takes place globally and locally

It will become obvious that the last approach is preferable in certain cases, such as in the next section, where we will orient our discussion of mesh control around solid elements.

Mesh control with three-dimensional elements

The only three-dimensional element within SOLIDWORKS Simulation is the solid element. We worked with this element in *Chapter 6, Analyses of Components with Solid Elements*, and *Chapter 7, Analyses of Components with Mixed Elements.*

As with the plane element demonstrated in the previous section, the idea of mesh control and refinement can also be applied when discretizing with solid elements. However, it is important to note that for large-scale structures, solid elements can quickly rack up degrees of freedom that will overwhelm your machine's memory. For this reason, you may find that the best compromise to arrive at a reliable result within a manageable computational time is to adopt the hybrid mesh refinement, where you combine a moderate global mesh with a locally refined mesh detail in the intricate regions. *Figure 10.13* shows the difference between the number of elements and nodes for a moderate mesh (*Figure 10.13a*), a moderate mesh with a local mesh refinement (*Figure 10.13b*), a fine mesh (*Figure 10.13c*), and a fine mesh with local mesh refinement (*Figure 10.13d*):

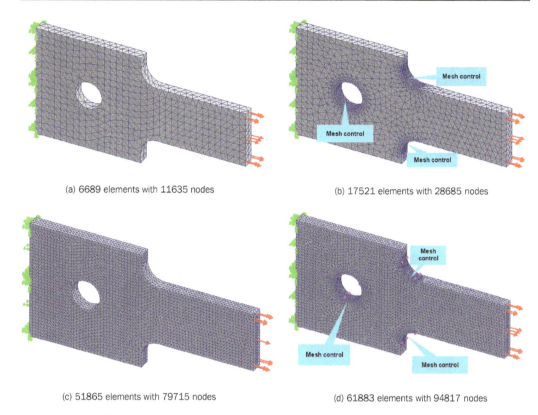

(a) 6689 elements with 11635 nodes

(b) 17521 elements with 28685 nodes

(c) 51865 elements with 79715 nodes

(d) 61883 elements with 94817 nodes

Figure 10.13 – An illustration of mesh refinements with solid elements

Let's see why a moderate global mesh with a locally refined mesh is a good compromise. You may recollect that a linear solid element has four nodes, as we mentioned in the *Characteristics of solid elements* subsection in *Chapter 7, Analyses of Components with Mixed Elements*. Each node has three **Degrees of Freedom (DoF)**. Thus, the total degrees of freedom to be solved by adopting the linear solid elements mesh type in *Figure 10.13b* is *28,685 x 4*, which equals 114632. Now, since finite element solutions involve matrix solutions, this means that behind the hood, the computer is solving a matrix sized 114,632 by 114,632. Now, compare this to the mesh detail in *Figure 10.13d* in which the total number of nodes is 94,817, resulting in a total of 379,268 DoF. For this mesh, the computer has to solve a matrix sized 379,268 by 379,268, which is more than three times the size of the matrix to be solved in *Figure 10.13b*. It is this increase in matrix size that contributes to a longer computational time and aggressive memory usage. Consequently, circumventing this means going with the approach of hybrid mesh control.

So, to wrap up the discussion, by adopting a moderate mesh size combined with a local mesh control, followed by a convergence study (as done in the previous section), you may be able to put some level of confidence in the accuracy of your results. There is one last remark – for very sharp corners or extremely small fillets, theoretically, you may expect an infinite value of stress (which is called stress singularity). However, it bears mentioning that there is no way to get an infinite value of stress from a finite element simulation. This is because you cannot have a mesh of size of zero! For instance, if you tried to key in zero for the mesh size parameter during meshing/mesh control, you will get an error message, as shown in *Figure 10.14*:

 Please enter a number between 1e-07 and 1e+06.

Figure 10.14 – An error from trying to enter a mesh of size zero during mesh control

This ends our discussion of meshing, mesh control, and mesh convergence study. In the last few pages, we have discussed the significance of these three concepts for one-dimensional elements (truss and beams), two-dimensional elements (plane elements), and three-dimensional elements (solid elements). Before we conclude the chapter, it may be of interest to know that apart from manually customizing the meshing detail to get accurate results, we can also employ a feature called adaptive solution methods in SOLIDWORKS Simulation. We've not covered this feature directly in the past; we shall briefly outline its essence in the next section to act as a source of further exploration for you.

Discretization with h- and p-type solution methods

The last concept we are going to call attention to is the adaptive solution method. For this, it is appropriate to begin with a brief highlight of the key parameters of the **Finite Element Method** (FEM). Technically, to obtain accurate solutions via finite element simulations, there are two methods of control that can be imposed to approximate the response (that is, displacement, stress, strain, and so on) of a structure:

- *Control of the mesh size*: As the name implies, this control is associated with reducing the mesh size, which often means that as we reduce the mesh size, and hence use more elements, it is expected that the solution will converge to a finite accurate solution.

- *Control of the order of polynomial*: At a fundamental level, the finite element method is a brilliant combination of the theories of polynomial approximation of differential equations and matrix analysis. Hence, controlling the order of the polynomial also tends to enhance the solution accuracy.

Altogether, the general idea is that in using polynomials to approximate a specific differential equation, we need to discretize the solution space (mesh), and then we can work with polynomials of different orders (low-order or higher-order). From this, it turns out that you can choose to find accurate solutions to a differential equation by any of the following means *[3-5]*:

- Fixing the mesh size while increasing the order of the approximation function, which is called the *p-version* of the FEM

- Fixing the order of the polynomial while decreasing the size of the mesh, which is known as the *h-version* of the FEM

- Using a concurrent variation of the mesh and the approximating polynomials, which is called the *hp-version* of the FEM

We've used the first two approaches indirectly in the past. For instance, at the beginning of *Chapter 6, Analyses of Components with Solid Elements*, we introduced the **Mesh** property manager, as shown in *Figure 10.15*:

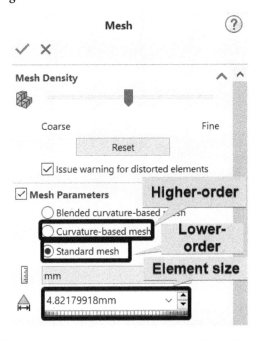

Figure 10.15 – The Mesh property manager for two- and three-dimensional elements

Essentially, switching between the **Standard** and **Curvature-based** mesh shown in *Figure 10.15* is analogous to increasing the order of the polynomial interpolating function. On the other hand, by decreasing the element size, we are varying the mesh size. However, beyond the aforementioned, another way to explore the *p-type* and *h-type* approaches is via the **Static Study** property manager shown in *Figure 10.16*. To obtain the **Static study** property manager shown in *Figure 10.16*, you should right-click on a static study's name and then select **Properties**. This should be done before discretizing a structure that is being analyzed.

Within the **Static** property manager, the **Adaptive** tab, wedged between the **Options** and **Flow/Thermal Effects** tabs that you have explored in the past chapters, presents a choice between an **h-adaptive** solution and a **p-adaptive** solution. Selecting any of the options will expose many more options to play with. The adaptive solution methods are algorithmic and automated ways of calling on SOLIDWORKS to create mesh refinement where needed, thereby saving users the need to make decisions about where mesh refinement should be applied:

Figure 10.16 – The adaptive solution method

For further usage of the adaptive solution method, a good place to start is the online help page from SOLIDWORKS: `https://help.solidworks.com/2021/English/SWConnected/cworks/IDC_HELP_PRESTATIC_P_ADAPTIVE.htm`.

Overall, there is growing literature on the adaptive approach to solution refinement, and it is worth exploring it in your journey toward the mastering of advanced tricks for finite element simulation of complex structures. For elegant coverage of the mathematical concept behind the adaptive solution approach to FEM, you can check out the book by *Szabó and Babuška* [6].

This concludes the coverage of the topics we set out at the beginning of this chapter. In effect, the ideas we have covered in the various sections of this chapter span the strategies that underpin the accuracy of finite element simulation results. Ultimately, the discussion of this chapter should complement other important lessons that stretch across the past chapters. By building on the discourse in this chapter and the knowledge accumulated from previous chapters, it is hoped that you have solidified your understanding of some of the various powerful analyses that can be done via finite element simulations. Congratulations!

Summary

Meshing is at the heart of the finite element simulation method. To get accurate results, you must rely on a high-quality mesh and the judicious use of mesh control. For this reason, we used the various sections of this chapter to offer a short overview of procedures for the mesh refinement of different elements. Overall, we did the following:

- We revisited the discretization of truss and beam structures to demonstrate how their mesh control can be deployed to reduce computational time.

- We documented the idea of the mesh convergence study and showed how it can be employed to facilitate the accuracy of results.

- We examined the benefit of mesh control for two-dimensional and three-dimensional elements and highlighted the procedure to strike balance between the accuracy of results and gaining computational efficiency.

As this summary marks the end of the book, it is worth examining how the book has progressed. In *Chapter 1, Getting Started with Finite Element Simulation*, we offered an overview of FEM and introduced the uniqueness of SOLIDWORKS Simulation for the analysis of engineering components. In *Chapter 2, Analyses of Bars and Trusses*, we commenced the analysis journey proper with the analysis of a crane whose parts were built from weldment profiles. We extended the scope of the analysis of components built with a weldment profile in *Chapter 3, Analyses of Beams and Frames*. Here, we demonstrated how to employ critical points along the length of beams to create appropriate line segments, highlighted the procedure to rotate a weldment profile, and suggested ways to apply distributed and bending moment loads on beams.

In *Chapter 4, Analyses of Torsionally Loaded Components*, we experimented with the creation of our first custom material, worked on the analysis of components with torsional loads, and revealed how to extract an angle of twists following the application of this load. *Chapter 5, Analyses of Axisymmetric Bodies*, initiated our treatment of advanced elements. Primarily, we covered the attributes of shell and axisymmetric plane elements and applied these elements to two case studies in the form of pressure vessels and a flywheel. In *Chapter 6, Analyses of Components with Solid Elements*, as the title implies, we shifted gear to the deployment of solid elements and worked on the analysis of helical spring and spur gears (coincidentally) as case studies.

With the major family of elements covered, we brought all of them together for the analysis of a multi-story building in *Chapter 7, Analyses of Components with Mixed Elements*. Via the multistory building example, we showed how to use the automatic contact pairs detection tool and highlighted how to employ the SOLIDWORKS Simulation in-built soft spring to provide stability. In *Chapter 8, Simulation of Components with Composite Materials*, we introduced the procedure for the analysis of components with composite materials. Thermal effects, cyclical load, and design study via optimization were the focal points of *Chapter 9, Simulation of Components under Thermo-Mechanical and Cyclic Loads*. And, finally, we ended it all with this chapter, where we have focused on meshing.

Although we've covered a lot of ground throughout the chapters, there is still so much that remains to be learned. The truth is, the field of finite element simulation is broad, and we have only scratched the surface. Nevertheless, it is hoped that the skills you've acquired will serve as a springboard to further exploration of advanced concepts.

Further reading

- [1] *Mechanics of Materials, R. C. Hibbeler, eBook, SI Edition, Pearson Education, 2017*

- [2] *A First Course in the Finite Element Method, D. L. Logan, Cengage Learning, 2011*

- [3] *The p-Version of the Finite Element Method, SIAM Journal on Numerical Analysis, Vol. 18, No. 3, I. Babuska, B. A. Szabo, and I. N. Katz, pp. 515–545, 1981*

- [4] *Interpolation and Quasi-Interpolation in h- and hp-Version Finite Element Spaces, Encyclopedia of Computational Mechanics Second Edition, T. Apel and J. M. Melenk, pp. 1–33, 2017*

- [5] *Bending of Microstructure-Dependent MicroBeams and Finite Element Implementations with R, R for Finite Element Analyses of Size-Dependent Microscale Structures, K. B. Mustapha, Springer, pp. 13–45, 2019*

- [6] *Finite Element Analysis: Method, Verification and Validation, B. Szabó and I. Babuška, 2021*

Index

S

T

`Packt.com`

Subscribe to our online digital library for full access to over 7,000 books and videos, as well as industry leading tools to help you plan your personal development and advance your career. For more information, please visit our website.

Why subscribe?

- Spend less time learning and more time coding with practical eBooks and Videos from over 4,000 industry professionals

- Improve your learning with Skill Plans built especially for you

- Get a free eBook or video every month

- Fully searchable for easy access to vital information

- Copy and paste, print, and bookmark content

Did you know that Packt offers eBook versions of every book published, with PDF and ePub files available? You can upgrade to the eBook version at `packt.com` and as a print book customer, you are entitled to a discount on the eBook copy. Get in touch with us at `customercare@packtpub.com` for more details.

At `www.packt.com`, you can also read a collection of free technical articles, sign up for a range of free newsletters, and receive exclusive discounts and offers on Packt books and eBooks.

Other Books You May Enjoy

If you enjoyed this book, you may be interested in these other books by Packt:

Practical Autodesk AutoCAD 2021 and AutoCAD LT 2021

Yasser Shoukry, Jaiprakash Pandey

ISBN: 978-1-78980-915-2

- Understand CAD fundamentals using AutoCAD's basic functions, navigation, and components
- Create complex 3d solid objects starting from the primitive shapes using the solid editing tools
- Working with reusable objects like Blocks and collaborating using xRef
- Explore some advanced features like external references and dynamic block
- Get to grips with surface and mesh modeling tools such as Fillet, Trim, and Extend
- Use the paper space layout in AutoCAD for creating professional plots for 2D and 3D models
- Convert your 2D drawings into 3D models

Learn SOLIDWORKS 2022

Tayseer Almattar

ISBN: 978-1-80107-309-7

- Understand the fundamentals of SOLIDWORKS and parametric modeling
- Create professional 2D sketches as bases for 3D models using simple and advanced modeling techniques
- Use SOLIDWORKS drawing tools to generate standard engineering drawings
- Evaluate mass properties and materials for designing parts and assemblies
- Join different parts together to form static and dynamic assemblies
- Discover expert tips and tricks to generate different part and assembly configurations for your mechanical designs

Packt is searching for authors like you

If you're interested in becoming an author for Packt, please visit `authors.packtpub.com` and apply today. We have worked with thousands of developers and tech professionals, just like you, to help them share their insight with the global tech community. You can make a general application, apply for a specific hot topic that we are recruiting an author for, or submit your own idea.

Hi!

I am Khameel B. Mustapha, author of *Practical Finite Element Simulations with SOLIDWORKS 2022*. I really hope you enjoyed reading this book and found it useful for increasing your productivity and efficiency in SOLIDWORKS Simulation.

It would really help me (and other potential readers!) if you could leave a review on Amazon sharing your thoughts on *Practical Finite Element Simulations with SOLIDWORKS 2022*.

Go to the link below or scan the QR code to leave your review:

`https://packt.link/r/1801819920`

Your review will help me to understand what's worked well in this book, and what could be improved upon for future editions, so it really is appreciated.

Best Wishes,